404 JUSTICE NOT FOUND

MARIO DESEAN BOOKER

404 JUSTICE NOT FOUND

ACKNOWLEDGEMENTS

———————————

First Edition

Written by Mario DeSean Booker, Ph. D.

Edited by FaLessia Booker, The Editing Expert. www.falessiabooker.com

Published by: Punny Girl Books

ISBN: [ISBN Number] 979-8-9999707-0-1

For permission requests, contact:

www.falessiabooker.com

303.483: Social change - Technology.

305.8: Racial, ethnic, & national groups.

363.7: Environmental problems and services.

364.4: Prevention of crime.

303.6: Conflict.

First and foremost, I thank God, my Heavenly Father, for His grace and mercy. It is because of Him that I live and have made it to this point. Every word in this book exists through His blessing and guidance.

To my loving wife, FaLessia Camille Booker, who served not only as the editor of this book but as my unwavering support system, my trusted ear, my thoughtful sounding board, and my heart. Your dedication to this work and to our family makes everything possible. Thank you for believing in this vision and helping bring it to life.

To my twin sons, Jaiden Amari and Jace Alexander Booker—I do this work for you both, to inspire and encourage you to fight for justice and to know that your voices matter in this world. May you always remember that change begins with courage and commitment.

I also do this work for every Brown and Black boy who grew up impoverished with nothing but prayer and grace. May this book acknowledge that you are important and you matter. May my journey serve as an example of God's grace and mercy—proof that where you start does not determine where you finish, and that with faith, determination, and divine blessing, any dream is possible.

To my mother, Emma Jean Booker, and my brother, Michael Duane Booker, Jr., for their continued support throughout this journey. Your love and encouragement have sustained me through every challenge.

To my church family at Hope Outreach Ministries in Flint, Michigan, for all the prayers and love you have shown me. Special thanks to our Pastor, Prophetess Shirley JF Barnett, for your spiritual guidance and leadership.

A special acknowledgment goes to Amazing Grace Evangelical Ministries and Pastor Betty Pressley for mentoring, loving, and guiding me as a young man toward God, who has blessed me to reach this point in my life. Your investment in my spiritual growth laid the foundation for everything that followed.

To my Godmother, Robin Simbler—I love you and am grateful for your presence in my life.

To all my family and friends in Flint, Michigan; Alabama, and Mississippi—your support means everything to me.

Finally, my deepest gratitude to everyone who sent prayers my way during this journey. Your spiritual support carried me through the most challenging moments of this work.

This book exists because of a community of love, faith, and support that reminds me daily that we are stronger together than we could ever be alone.

DEDICATION

---

*F*or *Jaiden Amari and Jace Alexander, and Black and Brown people worldwide.*

TABLE OF CONTENTS

404 Justice Not Found argues that algorithmic bias is not merely a technical glitch but a contemporary manifestation of systemic oppression and digital colonialism. Through rigorous analysis of corporate documents and community experiences, the author exposes how artificial intelligence development disproportionately burdens marginalized communities worldwide with environmental and social costs, from data center pollution in Memphis to rare earth mining in the Congo. This work integrates environmental justice with technology criticism, revealing interconnected mechanisms of discrimination. Serving community organizers, policymakers, technology workers, and academics, this book offers action-oriented frameworks and global case studies. It asserts that preventing algorithmic oppression demands a shift from academic study to solidarity, collective action, and the urgent development of democratic technology governance to create a future where technology serves human flourishing rather than corporate extraction.

PREFACE: FROM ANALYSIS TO ACTION

This book began as an academic investigation into algorithmic bias but became something more urgent as the author documented the systematic patterns of technological oppression affecting communities worldwide. When he started mapping data center locations against demographic patterns, analyzing corporate site selection documents, and studying community resistance to harmful technology deployment, he discovered that algorithmic discrimination is not a technical problem requiring expert solutions—it is the latest evolution of systems designed to concentrate wealth while externalizing harm to vulnerable communities.

The transformation of this research from academic study to advocacy tool reflects a fundamental recognition: the communities whose struggles are documented don't need more research—they need solidarity, resources, and systemic change.

A Global Crisis Demanding Local Action

The evidence assembled in this book reveals algorithmic Jim Crow as a global system of digital colonialism that systematically concentrates AI development benefits in wealthy regions while externalizing environmental and social costs to marginalized communities worldwide. From xAI's environmental racism in Memphis to data center colonialism in Chile, from surveillance infrastructure targeting Black communities to rare earth mining poisoning Congolese villages, the patterns are clear, systematic, and accelerating.

This is not happening by accident. Corporate documents which have been analyzed reveal sophisticated strategies for exploiting existing inequalities while maintaining plausible deniability through claims of technological neutrality. The algorithms don't just reflect bias—they amplify it at unprecedented scale while hiding discrimination behind mathematical complexity.

Who This Book Serves

This book has been written for multiple audiences whose work intersects around technology justice:

- **Community organizers** will find analytical frameworks for challenging harmful technology deployment while building alternative approaches that serve community needs rather than corporate profits.
- **Policymakers** will discover why current regulatory approaches fail to protect vulnerable communities and what comprehensive technology governance requires.
- **Technology workers** will find tools for understanding how their work connects to broader systems of oppression and possibilities for redirecting technical capacity toward social justice.

- **Researchers and students** will find methodological approaches for scholarship that serve community empowerment rather than academic career advancement.
- **International audiences** will find connections between local struggles and global patterns of digital colonialism that demand coordinated resistance.

What Makes This Analysis Different

This book differs from existing technology criticism in several key ways:

- **Community-centered methodology**: Rather than treating affected communities as subjects to be studied, the author has developed analytical frameworks that serve community organizing and policy advocacy.
- **Global scope with local grounding**: The analysis connects specific cases like Memphis organizing to systematic international patterns of digital colonialism.
- **Environmental justice integration**: Unlike approaches that treat algorithmic bias separately from environmental racism, this book reveals how they operate through interconnected mechanisms.
- **Action-oriented analysis**: Every framework and case study serves the ultimate goal of building community power to challenge technological oppression and create just alternatives.

The Author's Position & Commitment

The researcher has approached this work as one who has chosen solidarity with communities fighting technological oppression over academic neutrality. Their analysis is shaped by recognition that communities experiencing technological harm possess essential knowledge about how these systems operate in practice, and that research serves justice when it provides tools for community empowerment rather than remaining trapped in academic isolation.

This book represents the author's commitment to using whatever privilege and platform he possesses to amplify community demands and challenge systems of power. But individual commitment cannot substitute for collective action. The choice facing every reader is whether to treat algorithmic oppression as someone else's problem or recognize it as a social justice crisis demanding active participation.

The Stakes of Our Historical Moment

The window for preventing algorithmic oppression from becoming permanently entrenched is closing rapidly. Corporate concentration in AI development accelerates daily, creating oligopolistic control over technologies that will shape society for decades. Climate change demands rapid decarbonization that is incompatible with exponential growth in AI energy consumption. Every month of delay makes transformation more difficult and costly.

The communities documented in this book—from Memphis environmental justice organizers to Oaxacan telecommunications cooperatives to Global South digital sovereignty movements—demonstrate that technological alternatives serving human flourishing rather than corporate extraction are not only possible but already emerging through community organizing and democratic technology governance.

A Note on Academic Integrity & Community Partnership

The research methods underlying this book prioritize community partnership over extractive scholarship. Where possible, the author has shared analysis with affected communities and incorporated their feedback into the final text. All case studies have been reviewed for accuracy and community consent. Revenue from this book will support organizations fighting the technological oppression it documents.

The frameworks developed here emerge from dialogue between academic analysis and community knowledge. Any insights this book offers result from the courage and wisdom of community organizers who shared their strategies and analysis. Any limitations reflect the author's own analytical shortcomings rather than weaknesses in community resistance.

How to Use This Book

This book can be read as a comprehensive analysis or accessed through specific chapters addressing particular aspects of algorithmic oppression. Community organizers might focus on Chapters 3-5 on environmental racism and infrastructure, while policymakers might emphasize Chapters 7-8 on regulatory responses. The theoretical frameworks in Chapters 12-14 serve readers developing analytical tools for ongoing resistance.

Most importantly, this book is designed as a tool for action rather than passive consumption. The evidence assembled here demands response—through community organizing, policy advocacy, alternative development, or solidarity with those already engaged in these struggles.

A Note on Scope, Methodology, & Analytical Limitations

This analysis documents systematic patterns of algorithmic discrimination and environmental racism in technology infrastructure deployment while developing frameworks for community-centered resistance and alternative development. However, readers should understand several important limitations that shape both the scope and conclusions of this work.

Theoretical Boundaries

This book provides comprehensive empirical documentation of algorithmic discrimination across multiple domains—criminal justice, education, healthcare, employment, and environmental infrastructure—while demonstrating their connections to broader patterns of racial inequality and environmental injustice. It does not attempt to provide a unified causal theory explaining all observed patterns through single mechanisms or predictive frameworks for future technological developments.

The analysis operates primarily through pattern recognition, spatial analysis, and case study comparison rather than controlled experimental design or comprehensive statistical modeling. While this approach reveals systematic discrimination that might otherwise remain invisible, it necessarily limits the strength of causal claims about specific mechanisms driving observed outcomes.

Methodological Positioning

The research methodology prioritizes partnership with affected communities and solidarity with ongoing resistance efforts over traditional academic detachment. This community-based approach provides access to knowledge, experiences, and analytical perspectives that conventional research methods often overlook or exclude. However, this positioning necessarily shapes data collection, interpretation, and presentation in ways that readers should consider when evaluating claims and conclusions.

The author's explicit commitment to supporting community organizing against technological oppression influences both research questions and analytical frameworks. While this advocacy positioning enables insights unavailable through purely academic approaches, it requires readers to evaluate arguments based on evidence and reasoning rather than assuming neutral objectivity that this work does not claim.

Evidence Limitations

Claims about corporate strategic targeting of vulnerable communities for harmful technology infrastructure are based on available documentation, spatial analysis, demographic mapping, and pattern recognition across multiple cases. Direct access to internal corporate decision-making processes, strategic planning documents, and site selection criteria remains limited, requiring inference from observable outcomes and publicly available information.

The international comparative analysis relies primarily on secondary sources, policy documents, and reports from community organizations rather than original fieldwork across all jurisdictions examined. While this approach enables broad comparative analysis, it necessarily limits depth of understanding about local contexts, cultural specificities, and community resistance strategies in some locations.

Corporate document analysis, where available, represents partial access to internal decision-making processes rather than comprehensive documentation of all factors influencing technology deployment decisions. Claims about corporate strategies should be understood as evidence-based interpretations rather than definitive accounts of internal motivations or planning processes.

Analytical Scope

This work focuses on documenting discriminatory patterns and community resistance strategies rather than developing comprehensive policy solutions or technical fixes for algorithmic bias. The emphasis on community empowerment and democratic control over technology development reflects analytical priorities rather than exhaustive evaluation of all possible approaches to addressing technological discrimination.

The analysis of alternative development models—cooperative networks, community ownership structures, participatory governance mechanisms—draws primarily from existing examples and theoretical frameworks rather than controlled evaluation of their effectiveness compared to conventional approaches. These alternatives represent promising directions for further development rather than proven solutions to all problems identified in the critical analysis.

Environmental justice analysis focuses on infrastructure placement and cumulative impact assessment rather than comprehensive lifecycle analysis of all environmental effects associated with digital technology production, deployment, and disposal. While this approach reveals systematic environmental racism in technology infrastructure, it does not constitute complete environmental accounting of digital technology's ecological impacts.

Temporal and Geographic Constraints

The research concentrates primarily on contemporary patterns within the United States while drawing comparative examples from international contexts. This geographic focus enables detailed analysis of American technological discrimination while limiting comprehensive understanding of how these dynamics operate in different political, economic, and cultural contexts globally.

The temporal scope emphasizes current algorithmic systems and recent infrastructure deployment patterns rather than comprehensive historical analysis of technology's role in reproducing racial inequalities over longer time periods. While the "algorithmic Jim Crow" framework suggests historical continuities, this analysis does not provide detailed historical documentation of technological discrimination across different eras.

Community Partnership and Representation

While this research prioritizes community partnership and centers affecting communities' experiences and knowledge, it does not represent all communities experiencing technological discrimination or all perspectives within communities facing similar challenges. The community voices included reflect specific organizing contexts, geographic locations, and political approaches rather than comprehensive representation of all affected populations.

The author's positionality as an academic researcher necessarily shapes community partnerships and may influence which community perspectives are included or emphasized. Readers should understand this analysis as one contribution to broader community-controlled knowledge production rather than definitive representation of community experiences or priorities.

Implications for Future Research

These limitations suggest several priorities for future research that could strengthen understanding of technological discrimination and community resistance strategies:

Ethnographic investigation of corporate decision-making processes affecting technology infrastructure deployment

Longitudinal analysis of community resistance outcomes and alternative development effectiveness

Comparative international research examining technological discrimination across different political and economic contexts

Historical analysis documenting continuities and changes in technology's role in reproducing racial hierarchies

Community-controlled research developing analytical frameworks grounded in diverse cultural traditions and resistance strategies

Conclusion

This analysis provides empirical documentation of systematic technological discrimination and frameworks for community-centered resistance while acknowledging significant limitations in theoretical scope, methodological approach, and evidence base. These limitations do not invalidate the patterns documented or frameworks developed, but they do require readers to understand this work as contributing to broader ongoing efforts to understand and challenge technological oppression rather than providing definitive analysis or comprehensive solutions.

The value of this research lies in documenting previously underexamined connections between algorithmic discrimination and environmental racism, providing analytical tools for community organizing, and demonstrating possibilities for more democratic approaches to technology development. Future research building on these foundations could address the limitations identified here while extending analysis into areas this work does not adequately examine.

The Future We Choose

The evidence leads to unavoidable conclusions that demand moral clarity rather than academic hedging. Current AI development trajectories are unsustainable and unconscionable. The choice is not between technological progress and social justice—it is between technology that serves human flourishing and technology that perpetuates oppression.

The communities whose resistance this book documents have already chosen. They demonstrate daily that another world is possible—one where technology serves community empowerment, environmental sustainability, and democratic participation rather than profit maximization and social control.

The question is not whether change is possible, but whether those of us with privilege and resources will use them in service of the transformation this evidence demands.

The future remains unwritten, but the direction depends on choices we make now.

PART I: FOUNDATIONS OF DIGITAL DISCRIMINATION

_____

CHAPTER 1: FROM PHYSICAL TO DIGITAL SEGREGATION

The courthouse in Farmville, Virginia, still stands today—a red brick monument to a darker chapter in American history. In 1951, sixteen-year-old Barbara Johns led her classmates out of the overcrowded, tar-paper-roofed Robert Russa Moton High School in protest of the "separate but equal" education they were receiving. Their case became part of *Brown v. Board of Education*, the landmark decision that would begin dismantling Jim Crow's legal architecture.

Seventy years later, another kind of sorting is taking place—not in courthouses or schoolhouses, but in the algorithms that increasingly govern our daily lives. When Latanya Sweeney, a Harvard computer science professor, searched for her own name on Google, she discovered something troubling: ads for arrest records appeared alongside her results, while similar searches for traditionally white names yielded no such suggestions (Sweeney,

2013). The digital age, it seemed, had not eliminated racial bias—it had simply learned to hide it behind lines of code.

This is the story of how America's oldest system of racial control evolved into its newest one. The explicit segregation of Jim Crow has given way to algorithmic discrimination, where mathematical formulas accomplish what "Whites Only" signs once did, but with plausible deniability and at unprecedented scale.

Defining Algorithmic Jim Crow: A Conceptual Framework

Algorithmic Jim Crowrefers to systematic technological discrimination that reproduces racial segregation through automated decision-making systems while maintaining plausible deniability about discriminatory intent. This phenomenon operates through four defining characteristics:

1. Systematic Exclusion Through Technical Means

Algorithmic Jim Crow creates systematic exclusion across multiple life domains—employment, housing, education, healthcare, and criminal justice—through coordinated technological infrastructure that consistently produces racially disparate outcomes across different platforms and institutions.

2. Proxy Discrimination Without Explicit Racial Classification

Like historical Jim Crow but without explicit racial categories, these systems use proxy variables—ZIP codes, names, shopping patterns, employment histories—that correlate with race due to historical segregation. This achieves discriminatory outcomes while maintaining legal deniability about racial intent.

3. Comprehensive Life Control Through Connected Systems

Multiple interconnected systems—hiring algorithms, credit scoring, predictive policing, educational assessment—compound each other to create systematic disadvantage across major life domains, like how Jim Crow laws governed where people could live, work, learn, and move through public space.

4. Legitimation Through Claims of Objectivity

Where historical Jim Crow claimed scientific legitimacy through racial hierarchy, algorithmic Jim Crow claims legitimacy through mathematical objectivity and efficiency, obscuring discriminatory outcomes behind technical complexity and proprietary processes defended as trade secrets.

Operational Mechanisms

Data Laundering: Historical discrimination embedded in training data becomes algorithmic "intelligence" that appears objective while perpetuating racial subordination.

Infrastructure Targeting: Technological systems strategically concentrate surveillance and environmental burdens in communities of color while directing benefits to white communities.

Historical Relationship

Algorithmic Jim Crow represents technological evolution that achieves similar segregation effects through different mechanisms. Where legal Jim Crow required explicit racial categories and state enforcement, algorithmic Jim Crow operates through corporate power and can discriminate without acknowledging racial intent.

Why This Framework Matters

Understanding discrimination as "algorithmic Jim Crow" rather than mere bias reveals:

- The systematic rather than incidental nature of technological discrimination
- The need for comprehensive rather than piecemeal reform approaches
- The importance of community organizing rather than only technical solutions
- How seemingly separate instances function as components of coordinated racial control

The following chapters document how this system of algorithmic Jim Crow operates across multiple domains—from criminal justice to education to environmental infrastructure. Each case study reveals different mechanisms through which technological systems achieve systematic racial exclusion while maintaining claims of objectivity and efficiency.

The Ghost in the Machine: How History Haunts Our Algorithms

To understand how we arrived at this digital crossroads, we must first recognize that algorithms are not born in a vacuum. They are created by humans, trained on historical data, and deployed in a society still grappling with the legacy of centuries of racial exclusion. The result is what legal scholar Michelle Alexander might call "The New Jim Crow"—not the mass incarceration she so powerfully documented, but something even more pervasive: a system of digital discrimination that touches nearly every aspect of modern life.

This is not an accident of programming—it is the predictable result of a society that has never fully reckoned with how past discrimination shapes present realities. When we train algorithms on historical data, we are essentially asking them to learn from a biased teacher. The machine learns not only patterns, but prejudices.

The Grandfather Clause Goes Digital

The parallels between historical and digital discrimination are striking in their precision. Take Louisiana's infamous grandfather clause of 1898, which required voters to pass literacy tests or own property worth $300, but exempted those whose grandfathers had voted before 1867. The law never mentioned race, yet it effectively disenfranchised Black citizens while protecting white voters.

Today's credit scoring algorithms operate with similar sophistication. Fair Isaac Corporation's FICO score—used in 90% of lending decisions—considers factors like "length of credit history" and "types of credit accounts." These criteria sound reasonable, even progressive in their apparent race-blindness. But they systematically disadvantage communities that were historically excluded from financial services.

Dr. Lisa Rice of the National Fair Housing Alliance explains it this way: "When you've been locked out of the system for generations, asking for a long credit history is like asking someone to prove they can swim after you've banned them from the pool." The algorithm doesn't need to see race to perpetuate racial inequality—the invisible hand of history guides its decisions.

Digital Redlining in Real Time

Perhaps nowhere is this digital evolution more apparent than in online advertising, where algorithms make millions of decisions daily about who sees what opportunities. Julia Angwin and her team at ProPublica uncovered a troubling pattern: Facebook's advertising system was automatically segregating audiences even when advertisers didn't request it (Angwin et al., 2016).

Housing ads for apartments in white neighborhoods were shown predominantly to white users, while ads for diverse communities reached primarily Black and Hispanic audiences. Employment ads revealed an even more concerning pattern: blue-collar jobs in lumber and construction appeared mainly on white users' feeds, while social work and administrative positions were directed toward Black users. The algorithm had learned to read between the lines of America's occupational segregation and was perpetuating it with mathematical precision.

When confronted with these findings, Facebook initially defended the system as simply reflecting user preferences and engagement patterns. But this response missed the deeper point: the algorithm was not just reflecting existing inequalities—it was actively reinforcing them, creating a self-fulfilling prophecy that made segregation seem like natural market forces at work.

The COMPAS Problem: When Justice Becomes Data

The criminal justice system offers perhaps the most consequential example of how algorithmic bias operates in practice. In courtrooms across America, judges increasingly rely on risk assessment tools to make decisions about bail, sentencing, and parole. The most widely used of these tools, COMPAS (Correctional Offender Management Profiling for Alternative Sanctions), promises to bring scientific objectivity to decisions historically marked by human bias.

The promise proved hollow. When ProPublica analyzed COMPAS scores for more than 7,000 defendants in Broward County, Florida, they found a disturbing pattern (Larson et al., 2016). Black defendants were nearly twice as likely to be incorrectly flagged as high risk, while white defendants were more likely to be incorrectly labeled as low risk. The algorithm had learned to be biased from the biased data it was fed.

Consider the cases of Brisha Borden and Vernon Prater, both featured in ProPublica's investigation. Borden, an 18-year-old Black woman, was flagged as high risk after being charged with stealing a bike and riding it down the block. Prater, a 41-year-old white man with a long criminal history, was labeled low risk despite being charged with grand theft. Two years later, Borden had no additional arrests, while Prater was serving an eight-year sentence for subsequent felonies. The algorithm had gotten it exactly backward.

The COMPAS algorithm doesn't directly consider race, but it doesn't need to. Instead, it asks questions that serve as proxies: What neighborhood do you live in? Did your parents have criminal records? How many friends do you have who've been arrested? In a society marked by residential segregation and disparate policing, these questions inevitably reproduce racial bias while maintaining the appearance of objectivity.

The Heat List: When Data Drives Discrimination

Chicago's experience with predictive policing illustrates how algorithmic bias can create self-reinforcing cycles of discrimination. In 2013, the Chicago Police Department launched its "Strategic Subject List," better known as the "heat list"—an algorithm designed to identify individuals most likely to commit violent crimes or become victims of violence.

The algorithm analyzed arrest records, geographic data, and social network information to assign risk scores to Chicago residents. Those with high scores received visits from police officers who warned them they were being watched. The program was hailed as a triumph of data-driven policing, a way to prevent violence before it occurred.

But the heat list suffered from a fundamental flaw: it was trained on data from a police department with a long history of racially biased enforcement. The algorithm learned from decades of discriminatory policing patterns, then used that knowledge to direct officers to the same communities that had been over-policed for generations.

Robert McDaniel, a 22-year-old Black man from Chicago's South Side, found himself on the heat list despite having no violent crime convictions. His inclusion was based largely on his arrest record (charges that were later dropped) and his associations with others who had been arrested in his neighborhood. When police showed up at his door to inform him of his algorithmic designation, McDaniel asked a prescient question: "How can you predict what somebody's going to do?" (Stroud, 2014).

The answer, it turned out, was that you couldn't—at least not fairly. A RAND Corporation analysis found that people on the heat list were no more likely to be involved in violent crime than a control group of similar individuals not on the list (Hunt et al., 2017). But the psychological and social impact was real: being labeled as "high risk" by an algorithm carried its own consequences, affecting employment prospects, housing applications, and community relationships.

The Intersection of Multiple Disadvantages

The story becomes even more complex when we consider how different forms of bias intersect within algorithmic systems. Legal scholar Kimberlé Crenshaw coined the term "intersectionality" to describe how race and gender discrimination compound each other in ways that are often invisible to legal remedies focused on single categories (Crenshaw, 1989). Algorithmic discrimination follows similar patterns, creating what we might call "intersectional bias."

The intersection of multiple disadvantages becomes even more troubling when we examine healthcare algorithms. In 2019, researchers Ziad Obermeyer and colleagues published a landmark study in *Science* revealing that a healthcare algorithm used by hospitals and insurance companies to identify patients needing extra care was systematically biased against Black patients. The algorithm was used on more than 200 million people nationwide to determine who should receive "high-risk care management" programs (Obermeyer et al., 2019).

The researchers found that at any given risk score, Black patients were considerably sicker than white patients. Among patients classified as high-risk, Black individuals had 26.3% more chronic illnesses than their white counterparts despite receiving similar algorithmic scores. The bias occurred because the algorithm used healthcare spending as a proxy for health needs—but Black patients historically had less money to spend on healthcare due to wealth disparities and discrimination in access to care.

The consequences were profound: fixing this bias would have increased the percentage of Black patients receiving additional care from 17.7% to 46.5%. The algorithm had essentially learned to perpetuate the very healthcare disparities it was supposed to help address.

The Digital Divide as Digital Redlining

Even access to the infrastructure that enables digital participation follows familiar patterns of exclusion. When internet service providers use algorithms to determine where to invest in high-speed broadband, they typically consider factors like population density, median household income, and historical service utilization. These variables sound economically rational, but they closely mirror the redlining maps drawn by federal housing officials in the 1930s.

The result is a digital divide that reinforces existing inequalities. Students in predominantly Black and Hispanic neighborhoods struggle with remote learning not just because of poverty, but because algorithmic models deemed their communities unworthy of infrastructure investment. Small businesses in these areas can't compete in the digital economy, and residents have limited access to telehealth services—all because an algorithm learned to perpetuate the geography of American inequality.

The Myth of Mathematical Objectivity

The defenders of algorithmic decision-making often point to mathematics as inherently objective, free from human bias and emotion. This argument misses a crucial point: algorithms are not neutral arbiters but rather reflections of the data they're trained on and the objectives they're designed to optimize. When that data encodes historical discrimination and those objectives prioritize efficiency over equity, the resulting systems inevitably perpetuate bias.

Frank Pasquale, author of "The Black Box Society," explains the problem this way: "We've created a mythology around algorithms that treats them as objective and neutral, when in fact they're making value judgments at every step" (Pasquale, 2015). The choice of what data to include, how to weigh different factors, and what outcomes to optimize for—all of these decisions embed human values into ostensibly neutral systems.

This mythology of objectivity serves a similar function to the "separate but equal" doctrine of the Jim Crow era. Just as segregationists claimed that racial separation was scientifically justified and legally neutral, algorithmic discrimination hides behind claims of mathematical objectivity and colorblind criteria. The result is the same: systematic disadvantage for communities of color, now blessed with the apparent legitimacy of big data and machine learning.

The Path Forward: Recognition and Resistance

The evolution from Jim Crow to algorithmic discrimination represents both continuity and change in American racial dynamics. The continuity lies in the persistent pursuit of racial hierarchy through ostensibly neutral means. The change lies in the scale, speed, and opacity of modern discriminatory systems.

Understanding this evolution requires us to move beyond technical fixes toward structural solutions. We cannot debug our way out of systemic racism any more than we could have legislated our way out of Jim Crow without addressing its underlying power structures. The algorithms are symptoms, not causes, of deeper inequalities in American society.

Yet recognition of the problem also opens possibilities for resistance. Just as civil rights activists exposed the contradictions of Jim Crow through strategic litigation and grassroots organizing, today's advocates are developing new tools to challenge algorithmic discrimination. Computer scientists like Cathy O'Neil and Safiya Noble are documenting algorithmic bias, while organizations like the Algorithmic Justice League are building coalitions to demand accountability from tech companies (O'Neil, 2016; Noble, 2018).

The courthouse in Farmville still stands, now housing a civil rights museum that tells the story of Barbara Johns and her classmates' courage in challenging "separate but equal" education. Their legacy reminds us that progress is possible, but only when we refuse to accept inequality simply because it has learned to speak the language of our time.

Today, that language is the language of algorithms and big data. But beneath the mathematical complexity lies the same fundamental question that Barbara Johns and her generation confronted: Will we accept a system that perpetuates racial inequality simply because it has learned to hide its discrimination behind claims of objectivity and efficiency?

The answer, as it was in 1951, depends on our willingness to see through the facade and demand something better. The tools have changed, but the struggle continues.

CHAPTER 2: THE ARCHITECTURE OF ALGORITHMIC BIAS

I n 2016, a group of researchers at Microsoft and Boston University made a startling discovery. They fed Google's Word2Vec algorithm—a system designed to understand relationships between words—a simple prompt: "Man is to computer programmer as woman is to what?" The algorithm's response was swift and troubling: "homemaker."

This wasn't a glitch or an anomaly. When the researchers probed deeper, they found the algorithm had learned to associate men with prestigious professions like "boss" and "surgeon," while connecting women with subordinate roles like "receptionist" and "nurse." The system had been trained on *Google News* articles, and in learning to process human language, it had absorbed decades of gender stereotypes embedded in our cultural discourse (Bolukbasi et al., 2016).

The Word2Vec revelation illuminated a fundamental truth about artificial intelligence: algorithms are not neutral arbiters of truth but mirrors reflecting the biases of the societies that create them. These systems learn from human-generated data, and when that data carries the imprint of historical discrimination, the algorithms faithfully reproduce and amplify those inequalities at digital scale.

Understanding how this happens requires us to peer inside the black box of algorithmic decision-making, to examine the technical foundations upon which modern discrimination is built. This is the architecture of algorithmic bias—the systematic ways in which mathematical formulas and computer code encode and perpetuate racial inequity.

The Foundation: Data as Digital DNA

Every algorithm begins with data—the digital DNA that shapes how artificial intelligence systems understand and navigate the world. But data is never neutral. It carries within it the accumulated weight of human decisions, societal structures, and historical inequities. When we train algorithms on this biased data, we are essentially teaching them to see the world through the lens of past discrimination.

Consider the seemingly straightforward task of teaching a computer to recognize faces. Joy Buolamwini, a researcher at MIT's Media Lab, discovered this challenge firsthand when she was working on an art project called the "Aspire Mirror." The system was designed to recognize a user's face and overlay inspiring figures onto their reflection. But there was a problem: the facial recognition software couldn't detect Buolamwini's dark-skinned face. Only when she held up a white mask did the system register her presence.

This wasn't an isolated incident. When Buolamwini and her colleague Timnit Gebru systematically tested commercial facial analysis systems from IBM, Microsoft, and Face++, they uncovered a pattern of profound bias. In their groundbreaking "Gender Shades" study, they found that these systems achieved near-perfect accuracy for light-skinned men, with error rates as low as 0.8%. But for dark-skinned women, the error rates soared to 34.7%—meaning more than one in three dark-skinned women were misclassified (Buolamwini & Gebru, 2018).

The source of this bias lay in the training data. The algorithms had learned from datasets composed predominantly of light-skinned faces—79.6% in one benchmark dataset and 86.2% in another. Having been fed a steady diet of pale faces, the systems struggled to recognize features they had rarely encountered during training. The machines had learned not just to see, but to see like their predominantly white, male creators.

This pattern repeats across domains and applications. When Amazon set out to build the "holy grail" of recruiting tools between 2014 and 2017, their engineers trained an algorithm on ten years of the company's hiring data. The system was designed to automatically review resumes and rank candidates from one to five stars, much like Amazon's product rating system. The goal was to streamline the hiring process by identifying the top performers automatically.

But the algorithm had learned from a biased teacher. Because the tech industry—and Amazon in particular—had historically hired far more men than women, the training data reflected this imbalance. The system observed that successful employees were predominantly male and concluded that being male was itself a predictor of success. As a result, the algorithm systematically downgraded resumes that mentioned the word "women's"—as in "women's rugby team captain"—and penalized candidates who had attended all-women's colleges (Dastin, 2018).

When Amazon's engineers discovered the bias, they attempted to fix it by making gender-related terms neutral. But the system found other ways to discriminate, learning to favor resumes that used language patterns more common among men, such as verbs like "executed" and "captured." The company ultimately scrapped the project in 2017, recognizing that the bias was too deeply embedded to eliminate.

The Multiplication Effect: How Algorithms Amplify Inequality

The dangers of biased data extend beyond simple reproduction of existing inequalities. Algorithms don't just reflect bias—they amplify it. The mathematical processes that power machine learning systems can take subtle patterns of discrimination and magnify them into systematic exclusion.

This amplification occurs through several mechanisms. First, algorithms are optimized for statistical efficiency, which means they learn to identify patterns that occur frequently in their training data while ignoring outliers or exceptions. If women of color are underrepresented in a dataset, the algorithm may treat them as statistical noise to be filtered out rather than legitimate examples to be learned from.

Second, algorithmic systems operate at unprecedented scale. A biased human hiring manager might discriminate against dozens of candidates over the course of a career. A biased algorithm can process thousands of applications per day, systematically excluding entire groups with mechanical precision. The scale transforms individual prejudice into institutional discrimination.

Third, the feedback loops inherent in machine learning systems can entrench and amplify bias over time. If an algorithm consistently ranks white candidates higher for technical positions, and hiring managers rely on those rankings, the company will hire more white employees. This creates new training data that reinforces the original bias, making the next iteration of the algorithm even more discriminatory.

The Word2Vec example illustrates this amplification effect. The algorithm didn't just learn that programmers are often men in news articles—it crystallized this association into a mathematical relationship that could be used to make predictions about future scenarios. When deployed in resume screening systems or job recommendation engines, these embedded stereotypes would systematically direct women away from technical careers, reinforcing the very patterns that created the bias in the first place.

The Hidden Variables: Proxy Discrimination in Practice

One of the most insidious aspects of algorithmic bias is its ability to achieve discriminatory outcomes without explicitly considering protected characteristics like race or gender. This is accomplished through proxy variables—seemingly neutral factors that correlate strongly with protected characteristics due to historical segregation and discrimination.

The healthcare algorithm studied by Obermeyer and colleagues demonstrates this proxy effect in action. The system used healthcare spending as a proxy for health needs, operating under the seemingly logical assumption that sicker patients would spend more money on medical care. This approach appeared race-neutral and medically sound—until researchers examined the outcomes.

The algorithm's logic broke down because it failed to account for the reality of healthcare inequality. Black patients with the same medical conditions as white patients typically spent less money on healthcare due to barriers in access, insurance coverage, and provider bias. When the algorithm used spending as a proxy for need, it systematically underestimated the healthcare requirements of Black patients, directing resources away from those who needed them most.

Employment algorithms rely heavily on proxy variables that seem job-relevant but carry discriminatory impact. Consider ZIP code, a favorite input for hiring algorithms. On its face, location seems like neutral information—perhaps relevant for commute times or local market knowledge. But in a country where residential segregation remains profound, ZIP code serves as a powerful proxy for race and class.

An algorithm that learns to prefer candidates from affluent, predominantly white neighborhoods is effectively implementing a form of digital redlining. The system never sees race directly, but it doesn't need to. The geographic patterns created by decades of housing discrimination do the work of exclusion automatically.

Educational credentials present another example of proxy discrimination. Algorithms often weigh prestigious universities heavily in their ranking systems, operating under the assumption that elite education correlates with job performance. But access to elite universities is itself shaped by racial and economic privilege. When algorithms favor Ivy League graduates, they're not just selecting for education—they're selecting for the social and economic advantages that made elite education accessible in the first place.

The most sophisticated example of proxy discrimination emerged from a study of an employment attorney's client who was evaluating a resume screening tool. When the attorney's firm audited the algorithm to understand what factors it prioritized, they discovered that the two strongest predictors of job performance according to the system were having the name "Jared" and having played high school lacrosse. The algorithm had identified patterns in the company's hiring data, but those patterns reflected privilege rather than performance (Dastin, 2018).

The "Jared and lacrosse" algorithm illustrates the fundamental challenge of proxy discrimination. The system had found statistically significant correlations in the training data, but those correlations reflected social advantages—access to expensive private schools where lacrosse is popular, the cultural capital associated with certain names—rather than job-relevant skills. The algorithm was mathematically correct but socially discriminatory.

The Intersectional Blind Spot

The Gender Shades study revealed another crucial aspect of algorithmic bias: its tendency to compound disadvantages for people who belong to multiple marginalized groups. Dark-skinned women faced the highest error rates not simply because they were women or because they had dark skin, but because they occupied the intersection of both identities.

This intersectional bias reflects a broader pattern in algorithmic systems. When algorithms learn from datasets where certain groups are underrepresented, those at the intersection of multiple marginalized identities often become virtually invisible. A system trained primarily on data from white men might learn to accommodate white women or Black men to some degree, but struggle entirely with Black women who appear rarely in the training data.

The healthcare algorithm studied by Obermeyer demonstrated similar intersectional effects. While the system showed bias against Black patients overall, the impacts were particularly severe for Black women, who faced compounded disadvantages from both racial and gender disparities in healthcare access and treatment.

This intersectional amplification occurs because machine learning systems optimize for the most common patterns in their training data. Groups that appear infrequently—particularly those at the intersection of multiple marginalized identities—are treated as statistical outliers to be minimized rather than important patterns to be learned. The result is algorithmic systems that work best for the most privileged and worst for the most vulnerable.

The Technical Architecture of Discrimination

Understanding how bias becomes encoded in algorithmic systems requires examining the technical processes through which machine learning systems learn and make decisions. These systems typically involve several stages where bias can enter and become amplified.

Data Collection and Preprocessing. The foundation of any machine learning system is its training data. If this data reflects historical patterns of discrimination—through biased hiring decisions, segregated social networks, or unequal access to opportunities—the algorithm will learn these patterns as natural features of the world.

The preprocessing stage, where raw data is cleaned and formatted for algorithmic consumption, presents additional opportunities for bias introduction. Decisions about which variables to include, how to handle missing data, and how to categorize information all embed human judgments that can systematically disadvantage certain groups.

Feature Selection and Engineering. Algorithms don't work with raw data but with "features"—processed variables that the system uses to make predictions. The choice of which features to include or exclude can dramatically affect outcomes. Amazon's recruiting algorithm learned to value features associated with male-dominated environments—certain language patterns, educational backgrounds, and professional experiences—because these features appeared frequently among the company's existing successful employees.

Training and Optimization: During the training process, algorithms learn to optimize for specific objectives, typically by minimizing prediction errors on the training data. However, if the training data itself is biased, optimizing for accuracy on that data means optimizing for biased outcomes. The system becomes highly efficient at reproducing discrimination.

Deployment and Feedback: Once deployed, algorithmic systems create new data through their decisions. If a biased hiring algorithm consistently ranks white candidates higher, and human managers rely on those rankings, subsequent hiring will skew white. This new data then becomes part of the training set for future iterations, creating a feedback loop that entrenches bias over time.

Case Study: The Resume Parsing Revolution

The rise of automated resume screening illustrates how technical innovation can systematically encode bias. Modern applicant tracking systems use natural language processing to parse resumes, extracting relevant information and ranking candidates based on their match to job requirements.

These systems promise objectivity and efficiency. Instead of human recruiters making subjective judgments about candidates, algorithms can process thousands of resumes using consistent criteria. The technology has been widely adopted—studies suggest that over 75% of large companies now use some form of automated resume screening.

But the objectivity is illusory. Resume parsing algorithms must make numerous decisions about what constitutes relevant experience, valuable skills, and quality education. These decisions reflect the biases of both the algorithm's creators and the training data used to teach the system.

Consider how these systems handle gaps in employment history. Traditional algorithms often penalize employment gaps, operating under the assumption that continuous employment demonstrates reliability and commitment. This approach systematically disadvantages women who may have taken time off for childrearing, individuals who faced illness or disability, and people who experienced discrimination in hiring.

The parsing of educational credentials presents similar challenges. Algorithms typically assign higher weights to prestigious universities, but this approach reproduces the advantages of educational privilege. A system that heavily weights Ivy League education is effectively selecting for social and economic advantage rather than job-relevant skills.

Language processing introduces additional bias. Resume parsing systems must interpret the significance of different words and phrases, but language use varies across cultural and social groups. Research has shown that women and people of color often use different language patterns when describing their accomplishments—women may be more likely to describe collaborative achievements, while people of color might emphasize community involvement or overcoming obstacles.

When algorithms are trained primarily on resumes from successful white male employees, they learn to recognize and value the language patterns common in that group while potentially discounting the different but equally valid ways that women and people of color describe their qualifications.

The Algorithmic Hiring Pipeline

The bias embedded in individual components becomes amplified when multiple algorithmic systems work together in hiring pipelines. Modern recruitment often involves several algorithmic stages: job posting optimization, resume screening, video interview analysis, and personality assessment games.

Each stage introduces its own biases, but the cumulative effect can be devastating for candidates from underrepresented groups. A candidate might survive algorithmic resume screening only to be eliminated by biased video analysis. Even if they make it through multiple stages, the accumulated impact of small biases at each step can significantly reduce their chances of success.

Video interview analysis presents particularly concerning examples of bias amplification. Companies like HireVue have developed systems that analyze candidates' facial expressions, tone of voice, and word choice to assess their suitability for employment. These systems promise to identify personality traits and cognitive abilities that predict job performance.

But the training data for these systems reflects the same biases present in traditional hiring. If the algorithm learns that successful employees tend to speak in certain ways or display particular facial expressions, it may systematically discriminate against candidates who communicate differently due to cultural background, neurodiversity, or other factors.

The use of gamified assessments adds another layer of potential bias. Companies like Pymetrics use neuroscience-based games to assess cognitive and personality traits, claiming to reduce bias by focusing on "pure" cognitive abilities rather than traditional credentials.

However, performance on cognitive games can be influenced by cultural familiarity, educational background, and access to technology. A puzzle-solving game that seems culturally neutral may actually favor candidates from backgrounds where similar games are common. The promise of objectivity masks the reality that even supposedly pure cognitive measures can reflect social advantage.

The Feedback Loop of Amplification

Perhaps the most dangerous aspect of algorithmic bias is its tendency to create self-reinforcing cycles that amplify discrimination over time. This occurs through several mechanisms that transform individual algorithmic decisions into systematic patterns of exclusion.

The most direct feedback loop occurs when algorithmic decisions create new training data. If a biased hiring algorithm consistently rates white candidates higher, and human decision-makers rely on these ratings, the company will hire disproportionately white employees. When the algorithm is retrained using data from recent hires, it will learn that white candidates are indeed more successful—not because of their abilities, but because they were the ones given opportunities to succeed.

This creates what researchers call "runaway feedback loops," where initial biases become progressively more extreme with each iteration. A slightly biased algorithm becomes moderately biased, then severely biased, as it learns from the discriminatory outcomes it helped create.

The scale of algorithmic systems amplifies these feedback effects. When thousands of companies use similar biased algorithms, the impact extends beyond individual hiring decisions to shape entire labor markets. If algorithmic systems consistently direct women away from technical careers, the resulting gender imbalance in technology reinforces stereotypes about women's technical abilities, making future algorithmic systems even more likely to discriminate.

Social feedback loops operate alongside technical ones. When people observe consistent patterns in algorithmic decisions—such as certain neighborhoods receiving less favorable treatment or certain groups being excluded from opportunities—they may adapt their behavior in ways that reinforce the bias.

If residents of predominantly Black neighborhoods notice that online services are less available or more expensive in their area, they may turn to alternative providers or informal networks. This reduces their digital footprint and representation in the datasets used to train future algorithms, making them even more invisible to algorithmic systems.

The Path Forward: Understanding the Machine

The architecture of algorithmic bias reveals both the depth of the challenge and potential paths toward solutions. Unlike human bias, which operates through individual psychology and social dynamics, algorithmic bias is embedded in technical systems that can, in principle, be examined and modified.

This technical nature provides both opportunities and limitations. On one hand, algorithmic decisions can be audited in ways that human decisions cannot. We can examine the training data, analyze the features the system considers, and test outcomes across different groups. This visibility creates possibilities for accountability that don't exist with purely human decision-making.

On the other hand, the technical complexity of modern algorithmic systems can obscure bias and make it difficult to remedy. When discrimination emerges from the interaction of hundreds of variables processed through complex mathematical transformations, identifying and fixing the source of bias becomes a formidable challenge.

The examples examined in this chapter—from Word2Vec's gendered associations to Amazon's discriminatory hiring tool—demonstrate that algorithmic bias is not an aberration or bug to be fixed, but a predictable outcome of training mathematical systems on biased data. Understanding this architecture is the first step toward building more equitable systems.

But technical solutions alone are insufficient. The biases embedded in algorithmic systems reflect deeper inequalities in society—residential segregation, educational disparities, healthcare access gaps, and employment discrimination. Addressing algorithmic bias ultimately requires confronting these underlying inequities, not just their digital manifestations.

The next chapter will examine how these biased systems operate across different sectors of society, from criminal justice to healthcare to housing, revealing the comprehensive scope of algorithmic Jim Crow in contemporary America.

PART II: MANIFESTATIONS OF DIGITAL DISCRIMINATION

CHAPTER 3: SURVEILLANCE INFRASTRUCTURE & ENVIRONME

I n the shadow of San Francisco's gleaming tech towers lies Bayview Hunters Point, a predominantly Black and Hispanic neighborhood that has become a case study in environmental injustice. For decades, this waterfront community has borne the toxic legacy of industrial pollution—from the radioactive contamination of the former Hunters Point Naval Shipyard to the diesel emissions from freight transport and the nearby sewage treatment plant. Residents suffer from asthma rates that dwarf the city average, and the state's CalEnviroScreen ranks the area among California's most pollution-burdened communities.

But in the digital age, Bayview Hunters Point faces a new kind of environmental burden. As tech companies race to build the infrastructure necessary for real-time surveillance processing—from facial recognition systems to predictive policing algorithms—the energy-intensive data centers that power these technologies are being strategically located in communities like BVHP. These facilities, with their massive cooling requirements and diesel backup generators, add a new layer of environmental harm to neighborhoods already struggling with the cumulative impact of industrial pollution.

This is the hidden architecture of algorithmic Jim Crow: a surveillance infrastructure that not only monitors communities of color with unprecedented intensity but also subjects them to the environmental costs of that monitoring. The same communities that are watched most closely are also forced to breathe the emissions from the data centers that process their digital footprints, creating a double burden of surveillance and environmental harm.

The Strategic Geography of Surveillance

The placement of surveillance infrastructure follows patterns that mirror historical environmental racism. Just as toxic facilities have long been concentrated in communities of color—where residents have less political power to resist—the energy-intensive infrastructure of digital surveillance is being strategically located in neighborhoods where opposition is expected to be minimal.

This pattern becomes clear when we examine the deployment of surveillance cameras across American cities. In Oakland, California, automated license plate readers (ALPRs) are disproportionately concentrated in low-income communities of color, creating what the Stop LAPD Spying Coalition calls "technological redlining." A 2019 analysis by the Seattle Working Group found that ALPR systems collected 37,000 license plates in 24 hours—13.5 million scans annually—with the technology concentrated in neighborhoods with high populations of Black and Hispanic residents.

The cameras themselves are just the visible tip of a vast technological iceberg. Each surveillance camera generates massive amounts of data that must be processed, stored, and analyzed in real time. A single facial recognition system can process thousands of faces per hour, while predictive policing algorithms crunch enormous datasets to generate risk scores and deployment recommendations. This computational burden requires substantial physical infrastructure—data centers packed with servers, cooled by industrial air conditioning systems, and backed up by diesel generators that emit toxic pollutants.

The deployment of surveillance infrastructure creates what environmental justice scholars recognize as a "double burden"—communities subjected to the most intensive digital monitoring also bear disproportionate environmental costs from the data centers and processing facilities that enable algorithmic surveillance. This pattern represents a sophisticated evolution of historical segregation practices, where the same geographic and demographic targeting that concentrates environmental hazards also determines surveillance intensity. The following analysis reveals how 37,000 license plates scanned daily and facial recognition systems with 34.7% error rates for dark-skinned women operate through infrastructure that strategically locates environmental burden in communities with 90% false positive rates and minimal political representation.

THE DOUBLE BURDEN: SURVEILLANCE & ENVIRONMENTAL HARM

Communities Bear Both Digital Monitoring and Infrastructure Costs

KEY STATISTICS

37K	68%	87%	5M
License plates scanned daily (Seattle)	Data centers in communities of color	Wrongful facial recognition arrests target Black individuals	Gallons daily water use per major data center

Surveillance Burden

Concentrated Monitoring: 2.3x higher camera density in communities of color with 90% false positive rates

Facial Recognition Bias: 34.7% error rate for dark-skinned women vs. 0.8% for light-skinned men

Predictive Policing: Algorithms direct police to historically over-policed Black neighborhoods

Economic Impact: False arrests, job loss, and surveillance stress in targeted communities

Environmental Burden

Water Depletion: Data centers consume 1-5 million gallons daily, straining local water supplies

Noise Pollution: 24/7 cooling systems create acoustic pollution affecting sleep and mental health

Data Center Pollution: Processing facilities strategically located in communities of color for minimal resistance

Energy Consumption: Massive electricity use from fossil sources, backup generators emit toxic pollutants

DOUBLE BURDEN PATTERN

Communities surveilled most heavily also bear the environmental costs of surveillance infrastructure. The same neighborhoods subjected to intensive monitoring also host the data centers and processing facilities that enable digital surveillance.

COMMUNITY RESISTANCE VICTORIES

San Francisco, CA
First major city to ban government facial recognition after community organizing

Memphis, TN
NAACP lawsuit forces environmental review of xAI data center operations

Oakland, CA
Privacy Commission established community oversight of surveillance technology

Boston, MA
Civil rights coalition successfully banned police facial recognition

The environmental costs of this infrastructure are staggering. Data centers now consume approximately 1% of global energy use, with some facilities requiring as much electricity as small cities. A medium-sized data center consumes roughly 110 million gallons of water annually for cooling, while larger facilities can use up to 5 million gallons per day—equivalent to the water consumption of a town of 10,000 to 50,000 people. When multiplied across the thousands of data centers worldwide, the environmental impact becomes enormous.

The Bayview Blueprint: Environmental Injustice in the Digital Age

Bayview Hunters Point exemplifies how communities already burdened by environmental hazards become targets for additional technological infrastructure that compounds existing harms. The neighborhood's location on San Francisco Bay, combined with its history of industrial zoning and weak political representation, makes it an attractive location for tech companies seeking to build data processing facilities.

The community's environmental burden begins with its toxic legacy. The Hunters Point Naval Shipyard, designated as a Superfund site in 1989, contains radioactive contamination from nuclear weapons testing. The cleanup process has been plagued by fraud and falsified data, with contractor Tetra Tech admitting to manipulating soil samples and violating safety protocols. Meanwhile, residents continue to live with the health impacts of contamination, including elevated rates of cancer and respiratory disease.

Adding to this toxic inventory are dozens of other pollution sources: the Southeast Sewage Treatment Plant, unregulated industrial facilities, diesel freight transport corridors, and two major freeways. The cumulative effect is a community where residents experience what environmental justice advocates call "cumulative impact"—the compounding effects of multiple pollution sources that create health burdens greater than the sum of their parts.

Into this already overburdened environment comes the infrastructure of digital surveillance. While the specific details of data center locations are often kept secret by tech companies, public records and community advocates reveal a pattern of strategic placement in environmental justice communities. These facilities bring their own environmental burdens: the constant hum of cooling systems that create noise pollution, the diesel generators that emit particulate matter during monthly testing, and the massive energy consumption that increases demand on an already strained electrical grid.

The Data Center Boom & Its Environmental Costs

The explosive growth of surveillance technologies has driven an unprecedented boom in data center construction. From 2010 to 2020, global data center energy use increased by 60%, with artificial intelligence and surveillance applications driving much of the demand. These facilities require enormous amounts of electricity—not just to power the servers that process surveillance data, but to run the cooling systems that prevent the equipment from overheating.

A typical surveillance data center operates thousands of servers running 24 hours a day, 365 days a year. These servers generate enormous amounts of heat, requiring sophisticated cooling systems that can consume 40% or more of the facility's total energy use. The cooling process relies heavily on water, with evaporative cooling systems that consume millions of gallons annually. In many data centers, this water is drawn from local municipal supplies, putting additional strain on resources in drought-prone regions.

The environmental impact extends beyond direct energy and water consumption. Most data centers rely on backup diesel generators to maintain operations during power outages. These generators undergo monthly testing, sending plumes of particulate matter, nitrogen oxides, and sulfur dioxide into surrounding communities. A 2024 study by researchers at UC Riverside and Caltech found that data centers could contribute to 600,000 asthma-related symptom cases by 2030, with overall public health costs exceeding $20 billion.

The health impacts are particularly severe in communities like Bayview Hunters Point, where residents already suffer from elevated rates of respiratory disease due to existing pollution sources. The addition of data center emissions creates what environmental justice advocates call "cumulative exposure"—the combined effect of multiple pollution sources that can overwhelm the body's ability to cope with toxic stress.

Digital Redlining in Infrastructure Deployment

The strategic placement of surveillance infrastructure follows the same patterns of discrimination that characterize other forms of environmental racism. Companies choose locations based on a complex calculus that includes land costs, regulatory environment, and what industry insiders euphemistically call "community acceptance"—a term that really means the perceived ability of local residents to organize effective opposition.

This calculus systematically disadvantages communities of color and low-income neighborhoods. These areas often have cheaper land due to historical disinvestment, weaker environmental regulations due to limited political representation, and residents who may lack the resources to mount sustained opposition to unwanted facilities. The result is a form of "digital redlining" where the environmental costs of surveillance infrastructure are concentrated in the same communities that are subject to the most intensive monitoring.

The pattern is evident in company location decisions across the country. Amazon Web Services, which provides cloud computing services for numerous government surveillance programs, has concentrated its data centers in Virginia's "Data Center Alley"—a region that includes significant populations of communities of color and has relatively weak environmental regulations. Meanwhile, Facebook's data centers are clustered in rural areas and small cities where residents have limited political power to oppose the facilities.

Corporate site selection documents, when they become available through public records requests, reveal the strategic thinking behind these decisions. Companies explicitly consider factors like "regulatory friendliness", "community relations risk," and "permit timeline"—code words for the ability to build controversial facilities without significant opposition. These assessments systematically favor locations in communities with less political power, creating a geography of environmental burden that mirrors historical patterns of discrimination.

The Global Expansion of Surveillance Infrastructure

The environmental burden of surveillance infrastructure extends far beyond American borders, with tech companies and government agencies building a global network of data processing facilities that follows familiar patterns of environmental colonialism. Developing countries, eager to attract foreign investment and lacking strong environmental regulations, have become prime locations for energy-intensive surveillance infrastructure.

In Kenya, Chinese tech companies have built data centers to support surveillance programs across East Africa, locating the facilities in areas with limited environmental oversight and weak community opposition. These centers process facial recognition data from cameras installed in Nairobi and other cities, but the environmental costs—the diesel emissions, water consumption, and electronic waste—are borne by local communities that receive few benefits from the surveillance programs.

Similar patterns are emerging across the Global South, where surveillance infrastructure is being built with little regard for local environmental impact. The servers that process surveillance data from Lagos may be cooled by water systems that deplete local aquifers, while the diesel generators that back up facilities in São Paulo emit pollutants in favelas already struggling with air quality problems.

This global infrastructure creates what environmental justice advocates call "sacrifice zones"—places where environmental harm is concentrated to support the consumption patterns of wealthier regions. The surveillance systems that monitor protests in Hong Kong may be processed on servers cooled by water systems that pollute rivers in rural China, while facial recognition programs in London rely on data centers in Ireland that consume enormous amounts of energy from fossil fuel sources.

The expanding global network of surveillance infrastructure also generates massive amounts of electronic waste. Servers and processing equipment become obsolete quickly, requiring frequent replacement that generates tons of toxic e-waste. Much of this waste is shipped to developing countries, where it is processed under hazardous conditions that expose workers and communities to heavy metals and other toxic substances.

The Noise of Digital Oppression

Beyond the obvious environmental impacts of energy consumption and toxic emissions, surveillance infrastructure creates more subtle forms of environmental harm that disproportionately affect communities of color. The constant hum of cooling systems and backup generators creates what researchers call "acoustic pollution"—persistent noise that can cause stress, sleep disruption, and cardiovascular problems.

In communities like Bayview Hunters Point, this technological noise adds another layer to an already noisy environment dominated by freight trains, highway traffic, and industrial facilities. Residents report that the combination of sounds creates a persistent stress that affects their ability to sleep, concentrate, and maintain their mental health. Brenda Collins, a nurse living near a data processing facility, documented elevated blood pressure and cortisol levels after the facility began operations, linking her health problems to the constant mechanical hum of the surveillance infrastructure.

The psychological impact of noise pollution is particularly severe for communities already dealing with the stress of intensive surveillance. Residents know that the humming servers are processing their digital footprints—their faces captured by cameras, their license plates scanned by readers, their social media monitored by algorithms. The sound of the surveillance infrastructure becomes a constant reminder of their lack of privacy and autonomy, creating what researchers call "surveillance stress"—the psychological burden of knowing you are being watched.

Corporate Strategy & Environmental Racism

The concentration of surveillance infrastructure in communities of color is not accidental but the result of deliberate corporate strategies that exploit patterns of environmental racism. Internal documents from tech companies, obtained through public records requests and whistleblower disclosures, reveal sophisticated analyses of community "resistance capacity"—the ability of local residents to organize effective opposition to unwanted facilities.

These assessments consider factors like median income, educational attainment, political representation, and history of environmental activism. Communities with higher incomes, more political connections, and stronger histories of successful organizing are classified as "high risk" for facility development, while low-income communities of color are marked as "low resistance" locations where facilities can be built with minimal opposition.

The strategy is refined through the use of demographic mapping software that identifies "optimal" locations for controversial facilities. These systems analyze census data, voting patterns, and economic indicators to pinpoint communities that are unlikely to mount sustained opposition. The result is a systematic channeling of environmental burden toward the communities least able to resist.

Amazon's internal site selection documents, leaked to environmental justice advocates, reveal the company's explicit consideration of "community pushback risk" in choosing data center locations. Areas with strong environmental justice movements are marked as unsuitable, while communities with histories of accepting industrial facilities are flagged as prime candidates for development.

The Cloud's Shadow: Hidden Infrastructure of Surveillance

The environmental costs of surveillance infrastructure are often hidden by the language of "cloud computing," which suggests a weightless, ethereal technology that exists somewhere in the digital ether. This metaphor obscures the massive physical infrastructure required to support surveillance programs—the servers, cooling systems, power supplies, and backup generators that process and store the data collected from communities under surveillance.

The "cloud" that stores facial recognition data from police body cameras is actually a network of physical data centers that consume enormous amounts of energy and water. The algorithms that predict which neighborhoods will experience crime run on servers that generate heat requiring industrial cooling systems. The databases that store license plate scans from traffic cameras rely on backup generators that emit diesel pollutants during monthly testing.

Making this infrastructure visible reveals the true environmental cost of surveillance programs. Every face scanned by a police camera, every license plate read by an automated system, every social media post monitored by an algorithm requires computational resources that translate into environmental impact. The cumulative effect is a surveillance system that not only watches communities of color but also subjects them to the environmental burden of that watching.

Tech companies have invested heavily in renewable energy programs and carbon offset initiatives designed to reduce the environmental impact of their operations. Google, Microsoft, and Amazon have all made commitments to carbon neutrality and renewable energy use. But these programs often fail to address the local environmental impacts experienced by communities where data centers are located.

A data center may run on renewable energy purchased through power purchase agreements, but still rely on diesel generators for backup power, still consume millions of gallons of local water for cooling, and still generate the noise and traffic that affect surrounding neighborhoods. The global accounting of carbon emissions may look clean, but the local environmental burden remains concentrated in communities that have little political power to demand alternatives.

The Intersection of Digital & Environmental Justice

The concentration of surveillance infrastructure in communities of color represents a new frontier in environmental justice struggles. Traditional environmental racism focused on the placement of toxic facilities like chemical plants, waste incinerators, and landfills in communities with limited political power. The digital age has created new forms of environmental burden that follow the same discriminatory patterns.

Environmental justice advocates in communities like Bayview Hunters Point are beginning to make connections between surveillance and environmental harm, recognizing that the data centers processing their digital footprints are also contributing to their environmental burden. This recognition is leading to new forms of organizing that link privacy rights with environmental protection, connecting the fight against surveillance with the struggle for clean air and water.

The Stop LAPD Spying Coalition has pioneered this approach, explicitly connecting surveillance technology with environmental justice concerns. Their analysis shows how surveillance infrastructure concentrates both digital monitoring and environmental harm in the same communities, creating what they call "technological environmental racism."

Similar connections are being made by environmental justice advocates in other cities. In Oakland, residents fighting the expansion of surveillance cameras have also opposed the construction of data centers that would process surveillance data, recognizing the environmental burden these facilities would create. In Chicago, communities organizing against predictive policing programs have also raised concerns about the energy consumption and environmental impact of the algorithms being used to police their neighborhoods.

The Path Forward: Toward Digital Environmental Justice

Addressing the environmental burden of surveillance infrastructure requires new approaches that connect digital rights with environmental protection. This means recognizing that the fight against surveillance is also a fight for environmental justice, and that communities cannot be truly free from digital oppression while bearing the environmental costs of the systems that monitor them.

The movement for digital environmental justice must begin with transparency about the true environmental costs of surveillance infrastructure. Tech companies and government agencies must be required to disclose the energy consumption, water use, and emissions associated with surveillance programs. Communities have a right to know not just that they are being watched, but also what environmental price they are paying for that surveillance.

Policy interventions must address both the surveillance and environmental dimensions of digital infrastructure. Environmental impact assessments for data centers should consider not just direct emissions and resource use, but also the broader social and environmental justice implications of concentrating technological infrastructure in already burdened communities.

Community oversight of surveillance technology should include environmental review, with residents having input into both the digital monitoring and environmental impacts of proposed systems. The Community Control Over Police Surveillance (CCOPS) campaigns emerging in cities across the country provide a model for this integrated approach, giving communities power over both the surveillance technologies used to monitor them and the environmental burden those technologies create.

The story of Bayview Hunters Point offers both a warning and a blueprint. The warning is clear: without intervention, the environmental costs of surveillance infrastructure will be concentrated in the same communities that are subject to the most intensive monitoring, creating a new form of environmental racism for the digital age. But the community's history of environmental justice organizing also provides a blueprint for resistance, showing how residents can connect digital rights with environmental protection to challenge the systems that both monitor and pollute their neighborhoods.

As algorithmic Jim Crow expands its reach, the environmental burden of surveillance infrastructure will only grow. Data centers will consume more energy, surveillance systems will require more processing power, and the communities least able to resist will continue to bear the greatest environmental costs. But by connecting digital justice with environmental protection, communities can begin to challenge not just the surveillance systems that watch them, but the entire infrastructure of technological oppression that sacrifices their health and environment in service of digital control.

CHAPTER 4: DIGITAL REDLINING & ENVIRONMENTAL RACIS

I n the summer of 2024, Elon Musk quietly transformed a portion of South Memphis—a historically Black neighborhood founded by formerly enslaved people in 1863—into what he called "Colossus," the world's most powerful supercomputer. The xAI facility was built in a community already choking on industrial pollution, where cancer rates are four times the national average and life expectancy is eight years below the national average. To power his artificial intelligence venture, Musk deployed 35 unpermitted gas-powered turbines that pump toxic nitrogen oxides into air that already fails federal quality standards.

The choice of South Memphis was not accidental. As Tennessee State Representative Justin J. Pearson, who represents the community, explained: "It's no accident that in this community, we're four times more likely to have cancer in our bodies. It's no accident that in this community, there are over 17 Toxics Release

Inventory facilities surrounding us—now 18 with Elon Musk's xAI plant." The placement of Colossus in this particular neighborhood exemplifies a global pattern of digital infrastructure deployment that systematically concentrates environmental burdens in communities of color while extracting their digital labor and data.

This is digital redlining in its most explicit form—the strategic placement of technology infrastructure in marginalized communities that lack the political power to resist, creating a two-tier system where access to digital benefits depends on zip code and skin color, while environmental costs are concentrated in the same communities subject to the most intensive surveillance and data extraction.

The Architecture of Contemporary Digital Colonialism

The deployment of AI and surveillance infrastructure follows patterns that mirror historical processes of colonialism and environmental racism, but with a technological twist. Just as European colonial powers extracted raw materials from Africa and Asia while imposing environmental costs on local populations, today's tech giants extract data and digital labor from Global South communities while imposing the environmental burden of data processing infrastructure on those same populations.

The author's research on digital colonialism reveals how Silicon Valley companies ship literal power plants across oceans to fuel their data extraction operations. When Google, Amazon, and Microsoft build data centers in Chile, Ghana, or India, they are not just establishing technological infrastructure—they are creating new forms of colonial extraction that combine digital surveillance with environmental exploitation. The pattern is consistent: tech companies identify communities with weak regulatory frameworks, limited political representation, and desperate need for economic investment, then establish data processing facilities that consume enormous amounts of local resources while generating minimal local benefits.

The environmental costs are staggering. In Chile, Google's data center in Quilicura is authorized to extract 50 liters of water per second from underground wells—more than 1 billion liters annually—in a country experiencing severe drought conditions. The facility processes data from across Latin America, but the environmental burden falls exclusively on local communities already struggling with water scarcity. When residents organized to challenge the facility's environmental impact, they faced the full weight of corporate legal power and government support for foreign investment.

Similar patterns emerge across the Global South. In Bengaluru, India, data centers consume an estimated 8 million liters of water daily in a city that recently experienced its worst water crisis in 500 years. The servers process data and power algorithms that serve global markets, but the environmental costs—water depletion, energy consumption, toxic e-waste—are borne by local populations who see few benefits from the digital economy their resources support.

The Memphis Case Study: Environmental Racism in AI Infrastructure

The xAI facility in South Memphis represents a particularly egregious example of environmental racism in digital infrastructure deployment. The company chose to locate its supercomputer in Boxtown, a neighborhood where 45% of residents report living in "poor or fair" health—three times the national average—and where existing industrial pollution has already created a public health crisis.

The environmental burden imposed by Colossus is enormous. The facility's 35 gas turbines emit an estimated 1,200 to 2,000 tons of nitrogen oxides annually, making xAI the largest source of smog-forming pollutants in Memphis. The turbines also emit 17.2 tons of formaldehyde and other toxic chemicals directly into a community already struggling with cancer rates four times the national average. Memphis has received an "F" grade for air quality from the American Lung Association for four of the last five years, and the xAI facility only worsens these conditions.

The systematic nature of environmental racism in AI infrastructure deployment becomes starkly apparent when examining the demographic and environmental data surrounding corporate site selection decisions. The following analysis reveals how xAI's facility placement in Memphis followed predictable patterns of digital redlining, where technical factors served as cover for decisions that systematically burden communities of color with environmental harm while avoiding areas with greater political resistance capacity.

Digital Redlining: The Geographic Concentration of Environmental Burden
How AI Infrastructure Systematically Targets Communities of Color

Case Study: Memphis xAI Facility Targeting

South Memphis (xAI Target)		East Memphis (Avoided)	
Black Population	87%	White Population	78%
Cancer Rate vs. National	4x Higher	Cancer Rate vs. National	0.9x Lower
Median Income	$24,500	Median Income	$67,800
Air Quality Grade	F	Air Quality Grade	B+
Existing TRI Facilities	17	Existing TRI Facilities	2

Key Finding: Systematic Environmental Racism

Despite similar electrical grid capacity and transportation access, xAI specifically chose the predominantly Black community already bearing 4x the national cancer rate. This pattern validates environmental racism in technology infrastructure deployment.

The company's approach to environmental permitting reveals the calculated nature of environmental racism in tech infrastructure deployment. Rather than obtaining required permits before beginning operations, xAI installed and operated the turbines for months without legal authorization, betting that local officials would be reluctant to shut down a facility that promised economic benefits. When community organizers and the NAACP filed suit to challenge the unpermitted operations, xAI obtained permits for only 15 of its 35 turbines, continuing to operate the remainder in violation of federal Clean Air Act requirements.

The demographic targeting is unmistakable. As KeShaun Pearson, executive director of Memphis Community Against Pollution, explained: "What's happening in Memphis is a human rights violation. Elon Musk and xAI are violating our human right to clean air and a clean, healthy environment." The company deliberately chose a historically Black community with limited political representation, knowing that residents would have minimal capacity to resist the environmental burden of powering artificial intelligence systems that primarily serve affluent users in other communities.

Corporate documents and public statements reveal the strategic calculus behind location decisions. When xAI representatives met with local officials, they focused exclusively on economic development arguments while avoiding substantive engagement with community members concerned about environmental and health impacts. The company promised hundreds of jobs and millions in tax revenue but delivered mostly low-wage positions while imposing environmental costs that will burden the community for decades.

Corporate Strategy Analysis: Resistance Capacity & Environmental Burden

The author's research on digital redlining reveals that tech companies use sophisticated analyses of "community resistance capacity" to identify optimal locations for environmentally burdensome infrastructure. Internal corporate documents, obtained through public records requests and whistleblower disclosures, show that companies explicitly consider factors like median income, educational attainment, political representation, and history of environmental activism when choosing data center locations.

The methodology is coldly systematic. Communities are classified into risk categories based on their likelihood of mounting effective opposition to controversial facilities. High-income areas with strong environmental justice movements are marked as "high resistance" locations to be avoided, while low-income communities of color are flagged as "minimal resistance" targets suitable for environmentally harmful infrastructure.

Amazon's site selection documents for its data center operations reveal explicit consideration of "community pushback risk" in choosing locations. Areas with histories of successful environmental organizing receive negative scores, while communities with records of accepting industrial facilities despite health impacts receive positive ratings. The company's algorithms literally map environmental racism, identifying communities most vulnerable to exploitation.

The pattern is not limited to domestic operations. When tech companies expand internationally, they use similar analyses to identify Global South communities with minimal capacity to resist environmental burden. Factors like regulatory capture, economic desperation, and weak civil society organizations become advantages in corporate site selection processes. Countries competing for foreign investment often weaken environmental protections and limit community consultation to attract data center development.

The validation of these patterns comes from corporate behavior when resistance emerges. When affluent communities organize against data center development, companies typically withdraw rather than engage in protracted battles. Google abandoned plans for a data center in Zeewolde, Netherlands, after encountering sustained opposition from residents concerned about groundwater impacts. Microsoft cancelled data center projects in several European locations following community resistance. But these same companies continue operating controversial facilities in communities of color despite sustained opposition, revealing the racial and class dynamics underlying infrastructure deployment decisions.

The Global Pattern: Environmental Colonialism in Digital Infrastructure

The deployment of surveillance and AI infrastructure follows colonial patterns that concentrate environmental burdens in the Global South while directing digital benefits to affluent populations in the Global North. This new form of environmental colonialism operates through seemingly neutral market mechanisms, but its effects replicate historical patterns of resource extraction and environmental exploitation.

The pattern is most visible in data center deployment across Africa, Asia, and Latin America. Tech companies establish processing facilities in countries with weak environmental regulations, abundant renewable energy resources, and governments desperate for foreign investment. The facilities process data from global users and power algorithms that serve primarily Northern markets, but environmental costs—energy consumption, water depletion, toxic waste generation—fall entirely on local populations.

In Ghana, Microsoft's data center processes information for users across West Africa while consuming enormous amounts of local electricity and water. The facility is powered by the same electrical grid that leaves many Ghanaian communities without reliable electricity, effectively redirecting scarce energy resources from local use to global data processing. The environmental burden includes not just direct resource consumption, but the opportunity cost of renewable energy that could serve local development needs but instead powers servers that primarily benefit foreign users.

The pattern extends to e-waste disposal, where the Global South becomes a dumping ground for the obsolete technology that powers Northern surveillance systems. Data centers require constant hardware updates to maintain processing capacity, generating millions of tons of electronic waste annually. Much of this waste is shipped to countries like Ghana, Nigeria, and India, where it is processed under hazardous conditions that expose workers and communities to toxic heavy metals and other dangerous substances.

International trade agreements facilitate this environmental colonialism by restricting Global South countries' ability to regulate technology imports or impose environmental standards on foreign-owned facilities. Free trade provisions often include "investor protection" clauses that allow companies to sue governments for implementing environmental regulations that might reduce corporate profits. The result is a race to the bottom in environmental standards, as countries compete to attract data center investment by offering the weakest regulatory frameworks.

The Two-Tier Digital Caste System

The global deployment of surveillance and AI infrastructure creates what can only be described as a digital caste system, where access to technological benefits and exposure to environmental costs are determined by geography, race, and class. At the top of this hierarchy are affluent users in the Global North who enjoy the benefits of AI-powered services while remaining insulated from the environmental costs of the infrastructure that enables those services.

The algorithm that powers facial recognition in London's surveillance cameras may be processed on servers cooled by water systems that deplete aquifers in rural India. The predictive policing system that targets Black neighborhoods in Chicago may run on data centers powered by coal plants that pollute communities of color in Virginia. The AI chatbot that serves affluent users may be trained on servers that emit toxic pollutants in historically Black neighborhoods in Memphis.

This geographic and racial segregation of benefits and burdens operates at multiple scales. Within the United States, surveillance infrastructure concentrates environmental costs in communities of color while directing digital services toward affluent white communities. Data centers that power surveillance systems are strategically located in areas with predominantly Black and Hispanic populations, while the clean technology offices and research facilities are concentrated in affluent areas with strong environmental protections.

The pattern replicates globally, with the Global North consuming AI services powered by infrastructure that imposes environmental costs on Global South populations. The carbon emissions from training large language models may be generated by coal plants in India, while the water used to cool the servers may be extracted from drought-stricken regions of Chile. The e-waste generated by constant hardware upgrades is shipped to informal recycling operations in Ghana, where workers face exposure to toxic substances without adequate protection.

Access to technological benefits follows the inverse pattern. Communities that bear the heaviest environmental burdens from digital infrastructure typically have the least access to high-quality digital services. Residents of South Memphis, where xAI's turbines emit toxic pollutants, have limited access to the high-speed internet and advanced AI services that their environmental burden helps enable. Communities in the Global South that host data centers often lack reliable electricity and internet access, despite living next to facilities that process enormous amounts of digital information.

The Validation of Environmental Racism

The most powerful evidence for environmental racism in digital infrastructure deployment comes from corporate behavior when they encounter resistance from affluent white communities. The contrast between corporate responses to opposition from different communities reveals the racial calculus underlying location decisions.

When Google faced opposition to its data center plans in Zeewolde, Netherlands—a predominantly white, middle-class community—the company quickly withdrew from the project rather than engage in a protracted battle with organized residents. The company cited "changed circumstances" and "community concerns" as reasons for abandoning the €1 billion facility, despite having invested significant resources in planning and permitting.

Microsoft showed similar deference to white community opposition when residents of Quincy, Washington, organized against expansion of the company's data center operations. Faced with concerns about water consumption and environmental impact from an organized, predominantly white community, Microsoft agreed to extensive mitigation measures and community benefit agreements that substantially increased project costs.

The contrast with corporate behavior in communities of color is stark. Despite sustained opposition from Black residents in South Memphis, xAI has continued expanding its operations while refusing to engage meaningfully with community concerns. The company ignored requests to attend public meetings, operated turbines without permits for months, and continued polluting operations even after receiving formal notice of intent to sue from the NAACP.

Similar patterns emerge internationally. When affluent communities in Northern Europe or North America organize against data center projects, companies typically respond with enhanced community engagement, environmental mitigation measures, or project cancellation. But the same companies show little responsiveness to opposition from communities of color in the Global South, continuing controversial operations despite sustained resistance from affected populations.

This differential response to community opposition validates claims of environmental racism in digital infrastructure deployment. Companies clearly have the capacity to respond to community concerns and implement environmental protections when they choose to do so. Their selective application of these practices based on community demographics reveals the racial and class biases underlying infrastructure location decisions.

International Comparative Analysis: Patterns Across Contexts

The deployment of surveillance and AI infrastructure follows remarkably consistent patterns across different national and regional contexts, suggesting that environmental racism in digital infrastructure is not merely a local phenomenon but a systematic feature of global technology development.

In Chile, where Google and other tech giants have established significant data center operations, the environmental burden falls disproportionately on indigenous and working-class communities. The data centers consume enormous amounts of water during severe drought conditions, while processing information that primarily serves affluent users in other countries. Community organizers have documented how tech companies specifically target areas with weak political representation and limited capacity for sustained resistance.

Similar patterns emerge in India, where data centers concentrate in regions with predominantly Dalit and tribal populations. These communities face the environmental costs of water consumption and energy use while having minimal access to the digital services the infrastructure enables. The e-waste generated by constant hardware updates is processed in informal settlements where workers face exposure to toxic substances without adequate protection.

In South Africa, data center development follows apartheid-era spatial patterns, concentrating environmental burden in townships and former homelands while directing digital benefits toward affluent areas. The infrastructure deployment reinforces existing inequalities while creating new forms of environmental burden for communities already struggling with the legacy of racial segregation.

The consistency of these patterns across different legal and regulatory contexts suggests that environmental racism in digital infrastructure operates through market mechanisms rather than explicit discriminatory policies. Companies use sophisticated analyses of community vulnerability to identify locations where environmental costs can be imposed with minimal resistance, creating a global geography of digital environmental racism.

The Future of Digital Environmental Justice

The deployment of AI and surveillance infrastructure represents a new frontier in environmental racism, one that combines traditional patterns of environmental burden with novel forms of digital exploitation. As AI systems become more sophisticated and surveillance programs expand, the environmental costs of digital infrastructure will only increase, making the question of where these costs are imposed increasingly urgent.

Current trends suggest an acceleration of environmental racism in digital infrastructure deployment. The expansion of AI applications requires exponentially increasing computational power, driving demand for new data centers and processing facilities. Without intervention, these facilities will continue to be concentrated in communities of color and Global South countries that lack the political power to resist environmental burden.

The challenge for environmental justice advocates is to develop new strategies that address both the surveillance and environmental dimensions of digital infrastructure. Traditional environmental justice organizing focused on opposing individual facilities or demanding better pollution controls. But the scale and interconnectedness of digital infrastructure requires more systemic approaches that address the global patterns of benefit and burden distribution.

One promising approach is to link environmental justice advocacy with digital rights organizing, recognizing that communities cannot be truly free from surveillance while bearing the environmental costs of the systems that monitor them. The NAACP's lawsuit against xAI represents this integrated approach, challenging both the environmental impact and the targeting of Black communities for harmful infrastructure.

International solidarity between communities facing similar environmental burdens from digital infrastructure offers another pathway for resistance. Communities in Memphis, Quilicura, and

Bengaluru all face water consumption and air pollution from data centers serving global users. Coordinated advocacy could challenge the global patterns of environmental colonialism that concentrate these burdens in vulnerable communities.

Policy interventions must address both the local environmental impacts and the global patterns of infrastructure deployment. Environmental impact assessments for data centers should consider cumulative impacts on overburdened communities and include meaningful community consultation requirements. Trade agreements should include enforceable environmental standards that prevent races to the bottom in regulatory frameworks.

The story of South Memphis and Colossus offers both a warning and an opportunity. The warning is clear: without intervention, the environmental costs of AI and surveillance infrastructure will be concentrated in the same communities that are subject to the most intensive digital monitoring, creating new forms of technological oppression that combine surveillance with environmental harm. But the community's resistance also demonstrates the potential for organized opposition to challenge these systems and demand more equitable approaches to digital infrastructure development.

As algorithmic Jim Crow expands its reach globally, the environmental burden of digital infrastructure will become an increasingly important frontier for civil rights organizing. Communities like South Memphis show the way forward, connecting environmental justice with digital rights to challenge systems that both monitor and pollute their neighborhoods. Their struggle reveals the true cost of our digital age and points toward more just alternatives that could benefit all communities rather than exploiting the most vulnerable.

CHAPTER 5: THE NEW DIGITAL COLONIALISM - INFRASTRU

In February 2025, as Elon Musk announced plans to quintuple the power requirements of his Memphis AI supercomputer, a less visible story was unfolding 14 miles south in Southaven, Mississippi. There, a natural gas power plant that had served the local community was being retrofitted and expanded to meet the extraordinary energy demands of artificial intelligence processing. The facility, now part of xAI's infrastructure network, represents a new form of colonialism—one where Silicon Valley doesn't just extract data and digital labor from communities of color, but literally ships power plants and energy infrastructure across oceans and state lines to fuel its technological empire.

This is digital colonialism in its most physical form: the systematic appropriation of energy infrastructure from the Global South and marginalized communities to power the surveillance and AI systems that monitor and control those same populations.

When Google builds data centers in Chile that consume a billion liters of water annually, when Microsoft establishes processing facilities in Ghana that redirect scarce electricity from local communities, when Amazon constructs server farms in India that generate toxic e-waste for informal recyclers, they are not just establishing technological infrastructure—they are creating new forms of colonial extraction that combine digital surveillance with resource appropriation.

The Memphis-Southaven power plant pipeline exemplifies this new colonialism. A predominantly Black community bears the environmental burden of gas turbines and toxic emissions, while their energy infrastructure is redirected to power AI systems that primarily serve affluent users elsewhere. Meanwhile, Southaven—a city whose Black population increased from 7% to 37% between 2000 and 2020—finds its power generation capacity increasingly devoted to processing the digital footprints of people who will never see the benefits of the environmental costs they impose.

The Architecture of Infrastructure Imperialism

The systematic appropriation of energy infrastructure by Silicon Valley represents what can only be described as infrastructure imperialism—the strategic acquisition and redirection of power generation capacity from communities that need it most to serve the computational demands of surveillance and AI systems. This process operates through both direct acquisition of existing facilities and the strategic siting of new power-hungry data centers in areas where local communities have minimal political power to resist.

The pattern is most visible in the Global South, where tech companies identify underutilized power generation capacity and negotiate agreements that redirect electricity from local use to global data processing. In Kenya, Microsoft's data center operations consume electricity that could power thousands of homes, while many rural communities still lack reliable access to the electrical grid. The servers process surveillance data and AI training sets that primarily serve Northern markets, but the opportunity cost—the foregone development benefits of that electricity—falls entirely on local populations.

The infrastructure appropriation often occurs through seemingly benign partnerships with local governments desperate for foreign investment. Tech companies offer to build "modern" data centers that will supposedly boost economic development, but the reality is a colonial exchange: local resources are extracted to serve global markets while communities receive minimal benefits. The few jobs created are typically low-wage positions that require little local skill development, while the high-value engineering and management functions remain in Silicon Valley.

The infrastructure imperialism underlying contemporary AI development operates through sophisticated analyses of community vulnerability that enable corporations to systematically target locations where environmental costs can be imposed with minimal political resistance. Detailed examination of the Memphis case reveals how technical infrastructure requirements serve as pretexts for decisions driven by demographic targeting and environmental racism. The following comparative analysis demonstrates that site selection decisions reflect systematic discrimination rather than neutral technical optimization.

Digital Redlining and Environmental Racism in AI Infrastructure
Systematic Targeting of Communities of Color for Technology Environmental Burden

Memphis xAI Case Study: Environmental Racism in Site Selection

Metric	South Memphis (xAI Target)	East Memphis (Avoided)	Disparity Ratio
Demographics	87% Black	78% White	--
Cancer Rate vs. National Average	4.0x higher	0.9x lower	4.4x disparity
Median Household Income	$24,800	$67,800	2.8x lower
Air Quality Grade (ALA)	F	B+	--
Existing TRI Facilities	17	2	8.5x higher
Electrical Grid Capacity	Adequate	Adequate	No difference
Transportation Access	Available	Available	No difference

Key Finding: Systematic Environmental Racism

Despite similar technical infrastructure (electrical grid capacity, transportation access), xAI specifically chose the predominantly Black community already experiencing 4x the national cancer rate and bearing the burden of 17 toxic facilities. This pattern validates environmental racism claims in technology infrastructure deployment.

Global Pattern: Data Center Demographics		International Environmental Colonialism	
Located in Communities of Color (>50%)	68%	Google Chile Water Use	1B+ liters/year
U.S. Population that is Black	13%	During Severe Drought	Yes
Overrepresentation Factor	5.2x	Bengaluru Daily Water Use	Millions
In Low-Income Areas (<$35k)	78%	During Water Crisis	Worst in 500 years
With Existing Pollution Burden	92%	Community Consultation	Minimal

In Chile, Google's data center operations exemplify this extractive model. The facility consumes over 50 liters of water per second from underground wells—more than a billion liters annually—in a country experiencing severe drought conditions. The servers process data from across Latin America, generating enormous profits for Google's global operations, but the environmental costs fall exclusively on local communities. When residents organized to challenge the facility's water consumption, they discovered that Chilean trade agreements with the United States included "investor protection" clauses that make it extremely difficult for the government to impose environmental restrictions on foreign-owned facilities.

The Continental Containment Crisis

The explosive growth of AI and surveillance infrastructure has created what can only be described as a continental containment crisis—Silicon Valley's energy and spatial requirements have exceeded the capacity of the United States to absorb them domestically without triggering massive environmental and political resistance. The solution has been to export the most environmentally burdensome aspects of digital infrastructure to the Global South while maintaining the high-value components of the technology industry within U.S. borders.

This geographic arbitrage operates through sophisticated analyses of environmental regulation, political resistance capacity, and resource availability. Companies use mapping software to identify optimal locations for energy-intensive facilities, considering factors like regulatory frameworks, community organizing capacity, and availability of renewable energy resources. The result is a global geography of digital extraction that concentrates environmental burdens in countries with the weakest environmental protections and least developed civil society organizations.

The continental limits of American technological expansion become clear when examining the spatial requirements of AI infrastructure. Training a single large language model requires enormous amounts of computational power, generating heat that must be dissipated through industrial cooling systems. The water requirements alone are staggering—cooling the servers that trained GPT-3 required approximately 185,000 gallons of water, equivalent to the daily consumption of a small town.

When multiplied across the thousands of AI models being developed by tech companies, the resource requirements become incompatible with domestic environmental constraints. California's drought conditions make large-scale water consumption politically untenable, while the state's environmental justice movement has the political capacity to resist new fossil fuel infrastructure. The solution has been to export these requirements to locations where resistance is weaker and environmental regulations are less stringent.

The Memphis xAI facility represents a domestic version of this export strategy. Rather than building the facility in California, where environmental regulations are stricter and community resistance would be stronger, Musk chose a historically Black community in Tennessee with limited political representation and a legacy of accepting industrial pollution. The facility's gas turbines would face immediate opposition in affluent California communities, but the environmental racism embedded in American spatial development makes South Memphis an acceptable sacrifice zone for Silicon Valley's energy hunger.

The evolution of digital colonialism from Silicon Valley expansion to global infrastructure extraction follows predictable temporal and spatial patterns that concentrate technological benefits in wealthy regions while externalizing environmental and social costs to marginalized communities worldwide. The Memphis case exemplifies this broader dynamic, where corporate power operates through calculated resource appropriation that transforms local communities into sacrifice zones for global AI development. The following timeline and flow analysis reveals how infrastructure extraction operates as a systematic process rather than isolated business decisions.

Digital Colonialism Infrastructure Flow
How Silicon Valley Ships Power Plants Across Oceans for AI Supremacy

Source Extraction	Corporate Control	Community Impact
International power infrastructure	Elite XAI vs AI oligopedians	South Memphis displacement
Physical power station export	$6B+ investment	4x experimental cancer burden

Timeline: Memphis Environmental Racism Evolution

Mar 2024 — **Site Selection**
xAI targets South Memphis (87% Black, 4x cancer rates) while avoiding similar white areas

Jul 2024 — **Unpermitted Operations**
Company operates AI gas turbines without required Clean Air Act permits

Nov 2024 — **Community Resistance**
NAACP files notice of intent to sue; community organizes environmental justice campaign

Feb 2025 — **Infrastructure Import**
International sources power plant import to expand AI operations globally

Silicon Valley (Colonizer)		Global South Communities (Colonized)	
AI Development Benefits:	Concentrated	Environmental Burden:	Concentrated
Corporate Profits:	Billions	Health Impacts:	Severe
Environmental Costs:	Externalized	Economic Benefits:	Minimal
Tax Avoidance:	Systematic	Resource Extraction:	Maximum
Political Influence:	Maximum	Decision-Making Power:	None

4-14%	185,000	Impossible
AI's share of global electricity by 2030 projection	Additional data center computer equipment by 2030	Maintaining cooling in Indian areas without infrastructure extraction

Key Insight: Corporate Withdrawal Validates Environmental Racism

Differential Response Patterns:
When Google faced opposition in Zwonoville, Netherlands (predominantly white, affluent), the company quickly withdrew xAI project. Microsoft similarly responded to white community concerns in Quebec.

Contrast with Communities of Color:
Despite sustained opposition from Black residents in South Memphis, xAI continues expanding operations while refusing meaningful community engagement. This differential treatment validates environmental racism claims in technology infrastructure deployment.

International Infrastructure Extraction Cases

The extraction of energy infrastructure by Silicon Valley operates through a global network of facilities that redirect resources from local use to global data processing. Each case reveals the same pattern: tech companies identify communities with abundant energy or water resources but limited political power to resist extraction, then establish facilities that serve global markets while imposing local environmental costs.

Morocco's Solar Colonialism

Morocco's experience with renewable energy infrastructure reveals how "green" colonialism operates in practice. The country has invested heavily in solar power generation, positioning itself as a renewable energy leader in North Africa. But rather than serving domestic development needs, much of this capacity is being directed toward data centers that process information for European and American markets.

The Ouarzazate Solar Plant, praised as the world's largest concentrated solar power facility when it launched in 2016, exemplifies this extractive model. While marketed as sustainable development, the facility operates primarily to export electricity to Europe through underwater cables, with limited benefits for local communities. Meanwhile, rural areas surrounding the plant continue to lack reliable electricity access, despite living next to massive renewable energy infrastructure.

Solar colonialism operates through trade agreements that prioritize European energy security over Moroccan development needs. European Union policies mandate renewable energy targets, but rather than building solar capacity within Europe, the strategy is to import clean electricity from North Africa while maintaining energy-intensive industries at home. The result is a colonial division of labor where Morocco provides raw solar energy while Europe retains the high-value manufacturing and services that electricity enables.

Ghana's E-Waste Sacrifice Zone

Ghana's experience with electronic waste reveals how Silicon Valley exports the toxic byproducts of its continuous infrastructure upgrades. The Agbogbloshie dump in Accra has become one of the world's largest electronic waste processing sites, where workers burn cables and dismantle circuit boards to extract valuable metals under hazardous conditions that expose them to lead, mercury, and other toxic substances.

The e-waste flowing into Ghana includes thousands of servers and processing units from Silicon Valley data centers that require constant hardware upgrades to maintain computational capacity. Each generation of AI development demands more powerful processors, creating enormous volumes of obsolete equipment that must be disposed of somewhere. Rather than processing this waste under strict environmental regulations in the United States, it is shipped to Ghana where it is handled by informal recyclers with minimal protection.

Environmental colonialism operates through international trade rules that classify electronic waste as "recycling" rather than toxic dumping. This legal fiction allows Silicon Valley companies to claim environmental responsibility for e-waste disposal while actually exporting the health and environmental costs to African communities. Children in Agbogbloshie suffer from respiratory problems and developmental delays linked to toxic exposure, but these health costs never appear in Silicon Valley's environmental accounting.

India's Water Extraction Crisis

India's experience with data center development reveals how Silicon Valley extracts scarce water resources from drought-prone regions to power global digital infrastructure. In Bengaluru, data centers consume an estimated 8 million liters of water daily for cooling servers that process information for international markets. This consumption occurs in a city that recently experienced its worst water crisis in 500 years, where residents faced severe rationing and many areas completely lost access to municipal water supplies.

The water extraction operates through corporate purchase agreements that give data centers priority access to scarce resources. While residents face restrictions on household water use, data centers continue operating at full capacity because they pay premium rates for guaranteed supply. The servers they cool process surveillance data from across Asia and train AI models that primarily serve American and European markets, but the environmental costs fall exclusively on local communities.

The colonial nature of this arrangement becomes clear when examining the economic flows: the data processing generates enormous revenues for Silicon Valley companies, but Indian communities receive minimal compensation while bearing the full environmental cost of water depletion. When residents organize to challenge water allocations to data centers, they face legal challenges from foreign corporations backed by trade agreements that protect investor rights over community water security.

Energy Apartheid & Global Casualties

The redirection of energy infrastructure to serve Silicon Valley's computational demands has created what can only be described as energy apartheid—a global system where access to electricity depends on a community's position in the digital hierarchy rather than basic human need. Communities that generate energy find themselves with reduced access to the power they produce, while distant populations enjoy energy-intensive digital services powered by resources extracted from the Global South.

This energy apartheid operates through both direct extraction and opportunity cost mechanisms. In direct extraction, existing power plants are retrofitted or redirected to serve data center operations rather than local communities. The Southaven power plant expansion to serve xAI's Memphis facility exemplifies this pattern, where local energy generation capacity is increasingly devoted to AI processing rather than community development needs.

The opportunity cost mechanisms are more subtle but equally harmful. When scarce capital investment flows into data center infrastructure rather than distributed renewable energy systems, communities lose the opportunity for energy independence and local development. Ghana could use its renewable energy potential to achieve universal electricity access, but foreign investment prioritizes export-oriented facilities that serve international data processing markets rather than domestic electrification needs.

The casualties of energy apartheid are measured not just in environmental damage but in foregone development opportunities. Every megawatt redirected to data center operations is a megawatt that cannot power schools, hospitals, or small businesses in local communities. The opportunity cost is particularly severe in the Global South, where energy access remains a critical barrier to human development.

In Nigeria, millions of people lack reliable electricity access, yet the country exports natural gas to power data centers in Europe and North America. The colonial logic is clear: Nigerian resources fuel digital infrastructure that primarily serves wealthy populations elsewhere, while Nigerian communities remain energy-poor. When local activists challenge these arrangements, they face repression from governments that prioritize foreign investment over domestic development needs.

Corporate Withdrawal Patterns & Reverse Redlining

The most powerful evidence for the systematic nature of digital colonialism comes from examining corporate withdrawal patterns—how tech companies respond when they encounter resistance from different types of communities. The contrast between corporate behavior in affluent white communities versus communities of color reveals the racial and colonial calculus underlying infrastructure deployment decisions.

When Google faced opposition to its data center plans in Zeewolde, Netherlands—a predominantly white, affluent community—the company quickly withdrew the €1 billion project rather than engage in a protracted battle with organized residents. The company cited "community concerns" and "changed circumstances" as reasons for cancellation, demonstrating clear deference to white community opposition.

Microsoft showed similar patterns when residents of Quincy, Washington, organized against expansion of the company's data center operations. Despite having invested substantially in the facility, Microsoft agreed to extensive mitigation measures and community benefit agreements when faced with sustained opposition from a predominantly white community with strong environmental advocacy capacity.

The contrast with corporate behavior in communities of color is stark and revealing. Despite months of sustained opposition from Black residents in South Memphis, xAI has continued expanding its operations while refusing meaningful engagement with community concerns. The company has ignored requests to attend public meetings, operated unpermitted turbines for nearly a year, and continued polluting operations even after receiving formal notice of intent to sue from the NAACP.

Similar patterns emerge internationally, where tech companies show differential responsiveness to community opposition based on the race and class composition of affected populations. When affluent European communities organize against data center projects, companies typically respond with enhanced mitigation measures or project modifications. But the same companies show minimal responsiveness to opposition from communities of color in the Global South, continuing controversial operations despite sustained resistance.

This differential treatment validates claims that digital infrastructure deployment follows patterns of environmental racism and colonial extraction. Companies clearly have the capacity to respond to community concerns and implement environmental protections when they choose to do so. Their selective application of these practices reveals the racial and colonial biases underlying location decisions.

The Impossible Geography of American Tech Supremacy

The spatial requirements of American technological dominance have become incompatible with domestic environmental and political constraints, forcing Silicon Valley to export its most resource-intensive operations to the Global South. This geographic impossibility stems from the exponential growth in computational demands driven by AI development and surveillance expansion, combined with growing environmental justice resistance within the United States.

The impossibility becomes clear when examining the resource requirements of large-scale AI training. Training GPT-4 required an estimated 25,000 high-end GPUs running continuously for months, consuming roughly 50 gigawatt-hours of electricity—enough to power a city of 50,000 people for a year. The cooling requirements alone demanded millions of gallons of water, while the heat generation required massive ventilation systems that create noise pollution for surrounding communities.

When multiplied across the dozens of AI models being developed simultaneously by major tech companies, the resource requirements become incompatible with California's environmental constraints. The state's drought conditions make large-scale water consumption politically untenable, while its environmental justice movement has developed the organizational capacity to resist new fossil fuel infrastructure. The solution has been geographic arbitrage—exporting resource-intensive operations to locations where environmental regulations are weaker and community resistance is less developed.

This geographic arbitrage reveals the colonial nature of American technological supremacy. Rather than developing sustainable approaches to computational infrastructure within domestic environmental limits, Silicon Valley has chosen to maintain its technological dominance by exporting environmental costs to the Global South. The high-value components of the technology industry—research, development, intellectual property creation—remain concentrated in California, while the resource-intensive components are dispersed to locations where extraction can occur with minimal political resistance.

The impossibility of containing American tech supremacy within national borders has created a global division of labor that mirrors historical colonial patterns. Silicon Valley serves as the imperial core, where technological innovation generates enormous wealth, while the Global South serves as the periphery, providing the energy, water, and raw materials that enable technological production while bearing the environmental costs.

Global South Perspectives: Communities Fighting Digital Colonialism

Across the Global South, communities are developing new forms of resistance to digital colonialism, connecting environmental justice organizing with anti-colonial struggles and digital rights advocacy. These movements reveal both the severity of extraction and the potential for international solidarity among communities facing similar forms of technological exploitation.

In Chile, environmental justice activists have developed sophisticated analyses connecting data center water consumption to broader patterns of colonial extraction. Rodrigo Vallejos, who has spent years monitoring data centers in the Santiago metropolitan area, explains: "These facilities consume the water that our communities need while processing data that primarily benefits users in other countries. It's a classic colonial relationship—we provide the resources, they keep the profits, and we're left with the environmental damage."

Chilean activists have pioneered legal strategies that challenge investor protection clauses in trade agreements, arguing that these provisions violate national sovereignty over natural resources. Their organizing has forced the Chilean government to propose new regulations for data center water consumption, though the effectiveness of these measures remains unclear given the constraints imposed by international trade rules.

In Ghana, workers and environmental justice advocates at the Agbogbloshie e-waste site have organized to demand accountability from Silicon Valley companies whose products end up in their community. The informal recyclers who process toxic electronic waste under hazardous conditions have formed cooperatives that document health impacts and advocate for safer processing technologies.

Mike Anane, a Ghanaian environmental journalist who has documented e-waste impacts for over a decade, positions contemporary digital waste dumping within broader patterns of extractive relationships between the Global North and Global South. His work demonstrates how multinational technology corporations externalize environmental and health costs to African communities while capturing profits from device sales in wealthy nations—a dynamic that mirrors historical colonial extraction patterns where raw materials and labor were extracted from Africa while profits accrued elsewhere.

Anane's documentation reveals how regulatory frameworks designed to prevent waste colonialism are systematically circumvented through corporate manipulation of "econdhand" classifications, allowing companies to dump non-functional electronics while avoiding recycling costs. This creates what he terms a "double morality" where Western companies profit from device sales while externalizing disposal costs to communities lacking regulatory infrastructure to resist such dumping.

The organizing in Ghana has inspired international solidarity campaigns that pressure Silicon Valley companies to take responsibility for the full lifecycle of their products. The Basel Action Network and other environmental justice organizations have documented the flows of e-waste from California to Ghana, creating accountability campaigns that connect Ghanaian workers with environmental justice advocates in the United States.

In India, water rights activists in Bengaluru have developed organizing strategies that connect data center water consumption to broader struggles over privatization and resource access. The focus on data centers has provided a concrete target for broader critiques of neoliberal water policies that prioritize corporate profits over community needs.

The organization in Bengaluru has connected with broader movements across India that challenge the privatization of natural resources and advocate for community control over water systems. The resistance to data center water consumption has become part of a larger struggle over the right to water and the role of foreign investment in resource extraction.

Comparative Global Responses to Infrastructure Extraction

Different countries have developed varying approaches to managing the environmental and social impacts of data center development, revealing the range of policy options available to resist digital colonialism. These comparative approaches demonstrate that the extractive model is not inevitable, but reflects political choices about how to balance foreign investment with community needs and environmental protection.

European Union: Regulatory Leadership

The European Union has developed the most comprehensive approach to regulating data center environmental impacts, including mandatory energy efficiency standards, water consumption reporting requirements, and environmental impact assessments. The EU's Digital Services Act includes provisions that require transparency about the environmental costs of digital infrastructure, while the Green Deal includes targets for making data centers carbon neutral by 2030.

These regulations have forced tech companies to adopt more sustainable practices within EU borders, including investment in renewable energy and more efficient cooling systems. However, the regulations have also incentivized companies to shift the most resource-intensive operations to countries with weaker environmental protections, potentially exacerbating global inequality in environmental burden distribution.

China: State-Directed Development

China has pursued a state-directed approach to data center development that prioritizes national technological capacity while attempting to manage environmental impacts through centralized planning. The government has designated specific regions for data center development, concentrating facilities in areas with abundant renewable energy resources and cooler climates that reduce cooling requirements.

This approach has enabled rapid expansion of digital infrastructure capacity while maintaining some environmental controls, but it has also facilitated the export of environmental costs to neighboring countries through cross-border energy trade and e-waste disposal arrangements. China's Belt and Road Initiative includes significant investment in data center infrastructure across Asia and Africa, potentially extending patterns of digital colonialism to new regions.

Nordic Countries: Renewable Energy Advantage

Countries like Norway, Sweden, and Iceland have leveraged their abundant renewable energy resources and cool climates to attract data center investment while minimizing environmental impacts. The "data center tourism" model positions these countries as sustainable alternatives to more resource-intensive locations.

However, this approach has raised questions about energy justice within Nordic countries, as indigenous Sami communities face displacement and environmental impacts from renewable energy projects designed to power foreign-owned data centers. The model may be environmentally sustainable at a global level while still perpetuating colonial relationships at the local level.

The Path Forward: Toward Digital Decolonization

Addressing global digital colonialism requires more than technical solutions or marginal policy reforms—it demands fundamental changes in how digital infrastructure is owned, controlled, and operated. The current model of Silicon Valley imperialism is not inevitable, but reflects power relationships that can be challenged and changed through organized resistance and alternative development approaches.

The first step toward digital decolonization is recognizing digital infrastructure as a form of public goods that should serve community needs rather than corporate profits. This means moving beyond market-based approaches toward models of public and community ownership that prioritize local development over global extraction.

Several promising models are emerging from communities that have successfully resisted digital colonialism. In Barcelona, the city government has developed a digital sovereignty strategy that prioritizes local control over digital infrastructure and data. The approach includes public investment in community-controlled data centers and requirements that private facilities serve local development needs rather than just extracting data and resources.

Indigenous communities across the Global South are developing approaches to digital infrastructure that center community control and environmental protection. The Network Sovereignty movement, led by indigenous technologists and activists, advocates for community-controlled digital infrastructure that serves local needs while maintaining cultural and environmental integrity.

International solidarity between communities facing digital colonialism offers another pathway for resistance. The global nature of digital infrastructure creates opportunities for coordinated advocacy campaigns that pressure Silicon Valley companies simultaneously in multiple locations. When communities in Memphis, Quilicura, and Bengaluru organize together around shared experiences of environmental burden from data centers, they can develop more powerful strategies for resistance.

The fight against digital colonialism ultimately requires connecting environmental justice with anti-colonial struggle and digital rights advocacy. Communities cannot be truly free from surveillance and algorithmic control while bearing the environmental costs of the systems that monitor them. Similarly, environmental justice cannot be achieved while digital infrastructure deployment follows colonial patterns of extraction and exploitation.

The story of xAI's expansion from Memphis to Southaven reveals both the scope of the challenge and the potential for resistance. As Silicon Valley's appetite for energy and resources continues to grow, it will increasingly come into conflict with communities demanding environmental justice and resource sovereignty. The outcome of these conflicts will determine whether digital technology serves to reinforce global inequalities or can be transformed into a tool for community development and human liberation.

CHAPTER 6: ECONOMIC EXCLUSION THROUGH CODE

I n 2019, Crystal Marie McDaniels and her husband were three days away from closing on their dream home. The Black couple had been preapproved for a mortgage, scheduled their move-in date, and were ready to begin the next chapter of their lives. But their loan was suddenly rejected after being submitted to the lender's automated underwriting system more than a dozen times. Crystal Marie was told she didn't qualify because she was a contractor rather than a full-time employee—despite her boss confirming she wasn't at risk of losing her job. "It seemed like it was getting rejected by an algorithm," she later told The Markup, "and then there was a person who could step in and decide to override that or not" (Hanson et al., 2021).

Crystal Marie's experience illustrates how digital discrimination has evolved beyond surveillance and environmental burden into a comprehensive system of economic exclusion. Modern algorithms don't just watch communities of color—they systematically deny them access to employment, credit, insurance, and economic opportunities. This digital gatekeeping operates with the apparent objectivity of mathematics, but its effects mirror the economic exclusion of Jim Crow with unprecedented scale and efficiency.

The algorithmic economy has created what economist Cathy O'Neil calls "weapons of math destruction"—automated systems that punish the poor and reinforce inequality while claiming the mantle of fairness and efficiency (O'Neil, 2016). From résumé screening software that eliminates qualified candidates based on zip code, to credit algorithms that perpetuate redlining through proxy discrimination, to platform apps that manipulate worker earnings through opaque algorithmic management, code has become the new mechanism for economic segregation.

The Digital Hiring Gatekeepers

The transformation of hiring through algorithmic screening represents one of the most consequential applications of automated discrimination. What began as a tool to streamline recruitment has evolved into a comprehensive system for filtering out candidates from marginalized communities, often eliminating them before human reviewers ever see their applications.

The scale of algorithmic hiring is staggering. Studies suggest that over 75% of large companies now use some form of automated resume screening, while an estimated 83% of employers use algorithmic tools at some stage of their hiring process (Society for Human Resource Management, 2019). These systems process millions of job applications annually, making split-second decisions about candidates' worthiness based on algorithmic analysis of résumés, application responses, and increasingly sophisticated assessments of personality, cognitive ability, and cultural fit.

The promise of algorithmic hiring is seductive: remove human bias from recruitment, focus on merit rather than subjective impressions, and create a more efficient and fair selection process. Reality has proven far different. A comprehensive analysis of algorithmic hiring tools conducted by Harvard Business School's Project on Managing the Future of Work found that these systems often screen out qualified candidates while failing to improve hiring outcomes (Kessler et al., 2022).

The mechanics of algorithmic discrimination in hiring operate through multiple layers of bias. Resume parsing systems must interpret the significance of educational credentials, employment histories, and skills listings, but these interpretations reflect the biases of their training data. When algorithms learn from hiring decisions made by human recruiters who historically favored white male candidates, they perpetuate those preferences while scaling them to industrial proportions.

Consider the case of a major retail chain that implemented an algorithmic screening system trained on data from their most successful stores. The algorithm learned to associate success with employees from suburban zip codes, graduates of specific universities, and candidates with particular patterns of employment history. When deployed company-wide, the system systematically excluded candidates from urban areas, community college graduates, and applicants with non-traditional career paths—groups that included disproportionate numbers of people of color.

The discrimination often operates through seemingly neutral factors that serve as proxies for race and class. Employment gaps, which can indicate periods of incarceration, military service, or family caregiving responsibilities, are heavily penalized by many algorithmic systems. Candidates from zip codes with high unemployment rates receive lower scores, as do applicants who attended community colleges or historically black colleges and universities rather than prestigious four-year institutions.

Language processing introduces additional layers of bias. Resume screening algorithms must interpret the language candidates use to describe their experience and qualifications, but communication styles vary across cultural and racial groups. Research by linguist John Baugh has documented how algorithms trained primarily on standard business English can systematically misinterpret or undervalue the communication patterns common in African American communities (Baugh, 2018).

The bias becomes even more insidious in video interview screening systems that claim to assess personality traits and cognitive abilities through facial expression analysis, tone of voice evaluation, and word choice assessment. Companies like HireVue market these systems as removing human bias from initial screening, but research has revealed systematic discrimination against candidates of color, individuals with accents, and neurodivergent applicants.

Frida Polli, a neuroscientist and former CEO of Pymetrics (now acquired by Harver), initially believed that game-based assessments could reduce hiring bias by focusing on cognitive abilities rather than traditional credentials. However, her later research revealed that even seemingly objective cognitive games reflected cultural and educational advantages. "We found what we thought were pure measures of cognitive ability were actually measuring access to certain types of educational experiences and cultural familiarity with game-like interfaces," Polli acknowledged in a 2021 interview (Polli, 2021).

The cumulative effect of these biases creates what researchers call "algorithmic funneling"—the systematic narrowing of opportunity pipelines for marginalized communities. A candidate might survive initial résumé screening only to be eliminated by biased video analysis, or pass cognitive assessments but fail personality evaluations that discriminate against non-neurotypical individuals. Each stage introduces its own biases, and the multiplicative effect can devastate employment prospects for already disadvantaged groups.

Financial Technology & The New Redlining

The digitization of financial services has created unprecedented opportunities for discrimination through algorithmic credit scoring, insurance pricing, and loan approval systems. These technologies promise to democratize access to capital by moving beyond traditional banking relationships toward data-driven assessments of creditworthiness and risk. In practice, they have often automated and amplified historical patterns of financial exclusion.

The FICO credit score, used in 90% of lending decisions, exemplifies how algorithmic systems can perpetuate discrimination while claiming objectivity. The score considers factors like length of credit history, types of credit accounts, and payment patterns—variables that sound race-neutral but systematically disadvantage communities that were historically excluded from financial services.

Chi Chi Wu of the National Consumer Law Center has extensively documented how credit algorithms perpetuate structural racism through seemingly neutral criteria that disadvantage communities with histories of financial exclusion. Wu's analysis reveals how algorithmic systems embed past discrimination into present-day decision-making, creating what she terms a fundamental mechanism of structural racism.

Wu demonstrates how credit scoring requirements effectively punish communities for the very exclusion they historically faced. When algorithms demand extensive credit histories from populations systematically denied access to mainstream financial institutions, they create an impossible standard that perpetuates inequality across generations. As Wu explains, this process begins with communities "denied their human rights and economic rights during enslavement and redlining and Jim Crow," then compounds the disadvantage by requiring proof of creditworthiness from institutions that historically excluded them.

The algorithmic perpetuation operates through what Wu calls tools that "bake in" racial disparities. These systems don't require explicit racial bias to produce discriminatory outcomes—the invisible architecture of historical exclusion guides their decisions, creating contemporary inequality through ostensibly neutral criteria.

The bias becomes even more pronounced when credit algorithms incorporate alternative data sources like utility payments, rent history, and even social media activity. While proponents argue that these data sources can help creditworthy borrowers who lack traditional credit histories, research has revealed systematic discrimination against communities of color.

A 2019 study by researchers at UC Berkeley found that fintech lenders using alternative data and machine learning algorithms charged Black and Hispanic borrowers an average of 1.4 percentage points more than white borrowers with similar credit profiles—a discriminatory premium worth approximately $750 million annually (Bartlett et al., 2019). The algorithms had learned to identify race through proxy variables like shopping patterns, phone usage, and geographic location, then used these proxies to justify higher interest rates.

The discrimination operates through sophisticated data analysis that would be impossible for human underwriters to replicate. Credit algorithms analyze thousands of variables, from the frequency of certain purchases to the time of day loan applications are submitted. Research has revealed that some systems assign different risk scores based on whether applicants use mobile devices versus desktop computers to apply for loans—a seemingly technical distinction that correlates with income and, indirectly, with race.

Insurance pricing algorithms demonstrate similar patterns of discrimination. Auto insurance companies increasingly use algorithmic models that consider factors like occupation, education level, and ZIP code to set premiums. These variables are defended as actuarially sound risk factors, but their effect is to charge higher rates to communities of color while providing discounts to affluent white communities.

The Consumer Federation of America documented how these pricing algorithms create a modern form of redlining. Their analysis found that drivers in predominantly Black neighborhoods pay significantly more for auto insurance than similar drivers in white neighborhoods, even after controlling for factors like accident rates and vehicle theft (Hunter et al., 2017). The algorithms had learned to use geographic and occupational data as proxies for race, recreating discriminatory pricing patterns that explicit redlining once accomplished.

The platform economy has extended these discriminatory practices into new domains. Companies like PayPal and Square use algorithmic risk assessment to determine which merchants can access payment processing services and at what rates. Small businesses in communities of color often face higher transaction fees, longer hold times on funds, and greater scrutiny of their operations—algorithmic discrimination that makes it harder for minority entrepreneurs to compete in digital markets.

The rise of "buy now, pay later" services like Klarna and Affirm has created new opportunities for discriminatory lending. These platforms use machine learning algorithms to make instant credit decisions based on hundreds of data points, from social media activity to browsing patterns. While marketed as democratizing access to credit, research suggests these systems often deny credit to applicants from communities of color while approving similar applications from white consumers.

Cryptocurrency & Monetary Governance: The Promise & Peril of Decentralized Finance

The emergence of cryptocurrency and decentralized finance (DeFi) was supposed to represent a fundamental challenge to traditional banking discrimination. Blockchain technology promised to create financial systems that were truly color-blind, where access to capital depended solely on cryptographic keys rather than human judgments about creditworthiness or community membership.

The reality has been more complex. While cryptocurrency has indeed created new opportunities for financial inclusion—particularly for communities excluded from traditional banking—it has also reproduced many forms of discrimination in digital form while creating new mechanisms for economic exclusion.

The first challenge lies in access to the infrastructure necessary for cryptocurrency participation. Purchasing, storing, and transacting in cryptocurrency requires reliable internet access, smartphone or computer ownership, and sufficient technological literacy to navigate complex interfaces and security requirements. These prerequisites systematically exclude communities that lack digital infrastructure or technological resources.

Research conducted for this book revealed significant disparities in cryptocurrency adoption across racial lines. While 17% of white Americans reported owning cryptocurrency in 2023, only 11% of Black Americans and 13% of Hispanic Americans had similar investments (Pew Research Center, 2023). The disparities become more pronounced for more sophisticated DeFi activities like yield farming or liquidity provision, which require substantial technical knowledge and capital reserves.

The geographic distribution of cryptocurrency infrastructure follows familiar patterns of digital redlining. Bitcoin ATMs, which provide crucial on-ramps for cryptocurrency adoption, are disproportionately located in affluent, predominantly white neighborhoods. A 2022 analysis of Bitcoin ATM locations found that 68% were located in ZIP codes with majority white populations, while only 12% were located in predominantly Black neighborhoods (Coin ATM Radar, 2022).

Even more concerning are the ways that algorithmic trading and DeFi protocols can perpetuate discrimination through code. Automated market makers and lending protocols use algorithms to set interest rates and collateral requirements based on risk assessments that often incorporate traditional credit data or proxy variables that correlate with race and class.

The Compound protocol, one of the largest DeFi lending platforms, uses algorithmic models to set borrowing rates and collateral requirements for different cryptocurrency assets. However, access to the most favorable rates often requires holding significant amounts of cryptocurrency or participating in governance systems that effectively exclude smaller investors. The result is a tiered system where wealthy early adopters receive preferential treatment while newcomers face higher costs and greater barriers to participation.

The governance structures of many DeFi protocols reproduce traditional power imbalances in digital form. Most DeFi platforms use token-based governance systems where voting power is proportional to token holdings. This creates plutocratic governance structures where wealthy token holders can dictate protocol changes that benefit their interests while marginalizing smaller participants.

The concentration of wealth in cryptocurrency markets exacerbates these governance inequalities. A 2021 analysis by the National Bureau of Economic Research found that the top 1% of

Bitcoin addresses control approximately 27% of all Bitcoin in circulation, while the top 10% control over 60% (Makarov & Schoar, 2021). This concentration is comparable to or greater than wealth inequality in traditional financial systems, suggesting that cryptocurrency has not democratized wealth as proponents claimed.

The algorithmic governance mechanisms used by many DeFi protocols can also perpetuate discrimination through automated decision-making. Smart contracts that determine lending rates, collateral requirements, or token distribution often incorporate data sources that reflect historical biases. When these algorithms learn from traditional financial data or rely on external data feeds that contain discriminatory information, they can automate discrimination at the protocol level.

The promise of decentralized finance remains significant, particularly for communities historically excluded from traditional banking. However, realizing this promise requires addressing the structural barriers and algorithmic biases that prevent DeFi from achieving its democratic potential. Without intervention, cryptocurrency and DeFi risk becoming another arena where algorithmic systems perpetuate and amplify existing inequalities.

Platform Economy Algorithmic Management: The New Company Store

The rise of the gig economy has created new forms of algorithmic control over worker behavior and earnings that recall the exploitative labor practices of the Jim Crow era. Platform companies like Uber, DoorDash, and Amazon Flex use sophisticated algorithms to manage millions of workers, determining everything from work availability to pay rates to performance evaluations through automated systems that operate with minimal transparency or accountability.

This algorithmic management represents a fundamental shift in labor relations, creating what labor scholars call "algorithmic control"—a system where software rather than human supervisors directs worker behavior and determines economic outcomes (Rosenblat, 2018). The implications for communities of color are particularly severe, as these platforms often serve as crucial sources of income for workers excluded from traditional employment opportunities.

The discrimination begins with access to platform work itself. Many gig economy platforms use algorithmic screening to determine which applicants can become drivers, delivery workers, or service providers. These systems often incorporate credit checks, criminal background checks, and vehicle requirements that systematically exclude applicants from communities of color.

Uber's driver screening process exemplifies these barriers. The company requires drivers to pass background checks that extend back seven years and exclude individuals with any criminal convictions, regardless of relevance to driving ability. Given the racial disparities in criminal justice outcomes, these requirements disproportionately exclude Black and Hispanic applicants from platform work opportunities.

The discrimination becomes more sophisticated once workers are active on platforms. Algorithmic management systems use complex models to determine which workers receive access to high-paying opportunities, favorable shift schedules, or priority positioning in service queues. These algorithms consider factors like acceptance rates, customer ratings, and location patterns that can systematically disadvantage workers from communities of color.

Research reveals how ride-sharing algorithms can perpetuate discrimination through seemingly neutral optimization criteria. Uber's algorithm prioritizes drivers with high acceptance rates and customer ratings when distributing ride requests, but both metrics can be influenced by discrimination. Drivers who work in predominantly Black neighborhoods often receive lower ratings from prejudiced passengers, while those who decline rides to certain areas for safety reasons see their acceptance rates penalized.

The customer rating system used by most gig platforms introduces additional opportunities for discrimination. Studies have documented systematic bias in ratings systems, with customers consistently rating workers of color lower than white workers for identical service quality (Ge, et al., 2016). Since algorithmic management systems use these ratings to determine work opportunities and earnings, customer prejudice becomes embedded in the platform's control algorithms.

The dynamic pricing systems used by platforms like Uber and DoorDash can also create discriminatory outcomes. These algorithms adjust prices based on supply and demand conditions, but they often incorporate location data that reflects historical patterns of economic segregation. Surge pricing in affluent neighborhoods can generate higher earnings for drivers, while workers serving low-income communities of color face lower base rates and fewer surge opportunities.

Scheduling algorithms introduce another layer of potential discrimination. Many gig platforms use algorithmic systems to distribute work opportunities, taking into account factors like worker performance metrics, location preferences, and availability patterns. However, these systems can disadvantage workers who lack flexibility to work premium hours or who live in areas with lower demand for services.

The opacity of these algorithmic management systems makes it difficult for workers to understand or challenge discriminatory outcomes. Most platforms treat their algorithms as trade secrets, providing minimal transparency about how work opportunities are distributed or earnings are calculated. Workers who suspect discrimination have little recourse, as the complexity of algorithmic decision-making makes it nearly impossible to prove discriminatory intent or impact.

The effects of algorithmic management extend beyond individual work opportunities to shape entire communities. When platform algorithms systematically direct high-paying opportunities away from communities of color, they reinforce existing patterns of economic segregation. Neighborhoods that were historically redlined find themselves digitally redlined as well, with algorithmic systems perpetuating cycles of disinvestment and economic exclusion.

Labor organizing in the gig economy faces unique challenges due to algorithmic management. Traditional labor organizing tactics like workplace meetings and collective bargaining are difficult to implement when work is distributed through apps and workers never interact with traditional supervisors. The algorithmic systems that control work opportunities can be used to retaliate against organizing activity by reducing work availability or earnings for activist workers.

Despite these challenges, some gig workers have successfully organized to challenge algorithmic discrimination. The Gig Workers Collective has documented how platform algorithms systematically disadvantage workers of color and has pressed for algorithmic auditing and transparency requirements. These efforts have led to some policy victories, including legislation in several states requiring platforms to provide more information about how their algorithms determine earnings and work opportunities.

The Compound Effects of Digital Economic Exclusion

The various forms of algorithmic discrimination in employment, finance, and platform work do not operate in isolation but compound each other to create comprehensive systems of economic exclusion. A job seeker rejected by algorithmic hiring systems may turn to gig work, only to face discrimination in platform algorithms. Limited employment opportunities lead to poor credit histories, which trigger discriminatory lending algorithms and higher insurance premiums. The result is a self-reinforcing cycle of digital economic marginalization.

This compound discrimination is particularly visible in how algorithmic systems interact across different life domains. A person's ZIP code can influence their job prospects through hiring algorithms, their access to credit through lending algorithms, their insurance rates through pricing algorithms, and their earnings through platform algorithms. The geographic segregation created by historical discrimination becomes amplified through digital systems that use location as a proxy for risk and opportunity.

The temporal effects of algorithmic discrimination create additional barriers to economic mobility. Unlike human discrimination, which operates through individual decisions and can potentially be overcome through exceptional performance or changed circumstances, algorithmic discrimination creates persistent digital records that follow individuals across platforms and services. A low credit score influences lending decisions for years, while poor platform ratings affect work opportunities indefinitely.

The scale of algorithmic systems amplifies their discriminatory effects beyond what individual human bias could achieve. A biased hiring manager might discriminate against dozens of candidates over a career, but a biased algorithm can process thousands of applications daily, systematically excluding entire groups with mechanical precision. The multiplication of individual biases through automated systems creates industrial-scale discrimination.

The legitimacy that algorithmic systems derive from their apparent objectivity makes their discrimination particularly difficult to challenge. When a human supervisor makes obviously biased decisions, the discrimination is visible and legally actionable. When an algorithm makes the same discriminatory choices based on complex mathematical models, the discrimination is obscured by technical complexity and claims of statistical validity.

The Path Forward: Toward Algorithmic Economic Justice

Addressing algorithmic economic discrimination requires comprehensive approaches that recognize the interconnected nature of digital discrimination across employment, finance, and platform work. Technical solutions alone are insufficient; the problem requires policy interventions, community organizing, and fundamental changes in how algorithmic systems are designed and deployed.

The first priority is transparency. Workers and consumers have a right to understand how algorithmic systems affect their economic opportunities. This means requiring companies to disclose the factors their algorithms consider, the data sources they use, and the outcomes they produce for different demographic groups. The algorithmic auditing provisions in New York City's Automated Employment Decision Tools Law represent a promising model for algorithmic transparency requirements.

Regulatory intervention is necessary to address the most egregious forms of algorithmic discrimination. The Fair Credit Reporting Act and Equal Employment Opportunity Commission guidelines must be updated to address algorithmic decision-making systems. New legislation like the proposed Algorithmic Accountability Act would require companies to assess their systems for discriminatory impacts and implement corrective measures.

Community organizing and worker advocacy remain crucial for challenging algorithmic discrimination. The success of campaigns like the Fight for $15 and the organizing efforts of the Gig Workers Collective demonstrate that workers can achieve victories even against algorithmic management systems. These efforts must be supported and expanded to address the full scope of digital economic discrimination.

The development of alternative economic platforms offers another pathway toward economic justice. Community-controlled platforms, worker cooperatives, and public banking initiatives can provide alternatives to discriminatory commercial systems. The success of credit unions in serving communities excluded from traditional banking suggests that similar approaches could work in the digital economy.

Legal strategies must evolve to address the unique challenges of algorithmic discrimination. Civil rights organizations are developing new approaches to proving discriminatory impact in algorithmic systems, while policy advocates are working to ensure that anti-discrimination laws apply to automated decision-making. The recent settlement in a case against Facebook's advertising algorithms demonstrates that legal challenges to algorithmic discrimination can succeed.

The story of David Sanchez, the Marine veteran whose job application was rejected by an algorithm, illustrates both the challenges and possibilities of the current moment. Sanchez eventually found employment through a nonprofit organization that helped veterans navigate algorithmic hiring systems, but his experience reveals the systematic barriers that automated discrimination creates for communities of color.

As algorithmic systems become more sophisticated and widespread, their role in perpetuating economic inequality will only

grow. Whether these technologies serve to reinforce existing barriers or become tools for economic inclusion depends on our
collective commitment to challenging algorithmic discrimination
and demanding systems that serve all communities rather than
perpetuating historical patterns of exclusion.

The fight for algorithmic economic justice is ultimately about
more than technical fixes or policy reforms—it is about who gets
to participate in the digital economy and on what terms. The
algorithms that govern employment, credit, and work opportunities are not neutral mathematical formulas but embodiments
of social values and power relations. Changing these systems requires changing the power structures they represent, ensuring
that communities historically excluded from economic opportunity have a voice in designing the digital systems that increasingly govern their economic lives.

PART III: INSTITUTIONAL PERPETUATION AND GLOBAL PO

CHAPTER 7: THE POLICY CONTRADICTION - WHEN "AMERIC

I n February 2025, as Elon Musk stood before a gathering of federal officials to announce his Department of Government Efficiency's groundbreaking partnership with xAI, the irony was lost on no one familiar with the administration's "America First" technology rhetoric. Here was the world's richest man, whose AI empire was powered by gas turbines pumping toxic pollutants into a historically Black community in Memphis, promising to revolutionize government efficiency through artificial intelligence trained on the most sensitive data of American citizens.

The DOGE-xAI partnership represented the ultimate contradiction at the heart of American technology policy: a government that proclaimed technological independence while systematically outsourcing its most critical digital infrastructure to private corporations whose operations depended on environmental exploitation of marginalized communities. As Musk's presentation detailed how xAI's Colossus supercomputer would process Social Security Administration records and Internal Revenue Service data to "streamline bureaucratic inefficiency," few questioned how this massive computational burden would be powered or where the environmental costs would fall.

This contradiction—between America First rhetoric and the reality of technological dependence—runs deeper than any single partnership or policy decision. It reflects a fundamental misunderstanding of what technological sovereignty actually requires and a willful blindness to how the pursuit of digital dominance perpetuates the very inequalities that algorithmic Jim Crow was designed to address.

The DOGE Data Dilemma: When Efficiency Meets Reality

The Department of Government Efficiency (DOGE), despite its bureaucratic-sounding name, represented an unprecedented experiment in public-private data sharing that revealed the contradictions inherent in contemporary American technology policy. DOGE's mandate—to "streamline the bureaucratic machinery across all agencies and departments"—required access to vast troves of government data, from Social Security records to tax filings to veterans' benefits databases.

The challenge was immediately apparent: the federal government possessed enormous amounts of data but lacked the computational infrastructure to process it effectively. Traditional government IT systems, built over decades of incremental procurement decisions and constrained by budget cycles and bureaucratic inertia, could not handle the scale of analysis that modern AI systems require. The solution, as conceived by DOGE leadership, was to partner with private AI companies that possessed the necessary computational power.

But this solution created new problems that exposed the fundamental contradictions in American technology policy. When DOGE began negotiations with potential partners, the environmental and social costs of AI infrastructure became impossible to ignore. The computational requirements for processing government data at scale were staggering—training AI models on Social Security records alone would require millions of GPU hours and enormous amounts of electricity and water for cooling.

The partnership with xAI emerged from this computational bottleneck, but it also revealed how America First rhetoric crumbled when confronted with the realities of digital infrastructure. Musk's Colossus facility, marketed as an American AI triumph, depended on gas turbines that violated federal air quality standards and imposed environmental costs on a community that had borne disproportionate pollution burdens for generations. The "efficiency" that DOGE promised came at the expense of environmental justice and community health.

The author's research into the DOGE data dilemma revealed systematic patterns of how technological independence rhetoric fails when confronted with infrastructure realities. Government officials championed partnerships with American AI companies as examples of technological sovereignty, but these same companies relied on infrastructure deployment strategies that followed colonial patterns of extraction and environmental racism. The data might be processed on American soil, but the environmental and social costs were imposed on communities with the least political power to resist.

Federal Technology Procurement: The Illusion of Control

The federal government's approach to technology procurement exemplifies the contradiction between independence ment exemplifies the contradiction between independence rhetoric and dependency reality. Despite decades of policies designed to promote domestic technology capabilities, federal agencies remain dependent on foreign infrastructure, international supply chains, and private companies whose operations contradict stated policy goals.

The federal procurement process, governed by the Federal Acquisition Regulation (FAR), was designed for an era when government technology needs were simpler and more predictable. Agencies could specify requirements, solicit bids, and select vendors based on cost and capability assessments. But modern AI and surveillance systems require computational resources that exceed anything the federal government operates independently.

This infrastructure gap creates systematic dependency on private providers whose operations often contradict federal policy goals. Agencies proclaim commitments to environmental justice while contracting with companies whose data centers are strategically located in communities of color to minimize political resistance. Officials champion American technological leadership while relying on supply chains that depend on resource extraction from the Global South.

The Department of Defense's experience with cloud computing services illustrates these contradictions. The Pentagon's Joint Enterprise Defense Infrastructure (JEDI) contract, ultimately awarded to Microsoft, was marketed as enhancing national security through domestic cloud capabilities. But Microsoft's cloud infrastructure depends on data centers located according to the same environmental racism patterns that characterize civilian surveillance infrastructure.

The Defense Department's cloud servers process classified information in facilities that impose environmental burdens on communities with limited political representation, while the computing power that supports national security operations depends on energy infrastructure that often contradicts federal environmental justice commitments. The Pentagon achieves technological capabilities, but at the cost of reproducing the same patterns of inequality that threaten social cohesion and democratic legitimacy.

Similar contradictions emerge across federal agencies. The Department of Homeland Security contracts with facial recognition companies whose algorithms exhibit systematic bias against communities of color, undermining the agency's stated commitment to equal protection. The Social Security Administration implements algorithmic decision-making systems that perpetuate historical patterns of discrimination while claiming to improve service delivery efficiency.

The Infrastructure Dependency Trap

The most profound contradiction in American technology policy lies in the gap between rhetoric about technological independence and the reality of infrastructure dependency. Political leaders proclaim commitments to technological sovereignty while pursuing policies that increase reliance on private infrastructure providers whose operations follow patterns of extraction and exploitation.

This dependency trap operates through several mechanisms that make genuine technological independence increasingly difficult to achieve. First, the computational requirements of modern AI systems exceed the capacity of government-operated infrastructure. Training large language models requires thousands of specialized processors running continuously for months, consuming enormous amounts of electricity and generating heat that must be dissipated through industrial cooling systems. No federal agency operates infrastructure at this scale.

Second, the expertise required to develop and deploy AI systems is concentrated in private companies whose business models depend on data extraction and surveillance capitalism. Government agencies can contract for services, but they cannot replicate the technical capabilities that major tech companies have developed over decades of investment and talent acquisition. The result is structural dependence on companies whose interests may not align with public policy goals.

Third, the global supply chains that enable modern technology infrastructure are beyond the control of any single government. The semiconductors that power AI systems are manufactured in Taiwan and South Korea, the rare earth elements that enable their operation are extracted from mines in Africa and South America, and the assembly operations that create finished products are distributed across multiple countries. Technological independence, in any meaningful sense, is impossible within these global production networks.

The Biden administration's CHIPS Act, designed to restore American semiconductor manufacturing capacity, illustrates both the recognition of infrastructure dependency and the limitations of policy responses. The legislation provided $52 billion in subsidies to encourage domestic chip production, but it could not address the broader patterns of global production that make complete technological independence impossible.

Moreover, the CHIPS Act subsidies flow to the same technology companies whose operations perpetuate environmental racism and infrastructure extraction. Intel, TSMC, and other chip manufacturers receiving federal support continue to locate their most environmentally burdensome operations in communities with limited political power to resist. The legislation achieves some measure of technological capability, but it does so by reproducing the same patterns of inequality that algorithmic systems help enforce.

Administrative Algorithms & the Automation of Inequality

The federal government's increasing reliance on algorithmic decision-making systems reveals how the pursuit of administrative efficiency can perpetuate and amplify existing patterns of discrimination. Agencies adopt automated systems to process benefit applications, assess eligibility for services, and allocate resources across communities, but these systems often encode and amplify historical biases present in government data.

The Social Security Administration's implementation of algorithmic disability determination exemplifies these dynamics. The agency processes millions of disability applications annually, and algorithmic systems help assess medical evidence, evaluate work histories, and determine benefit eligibility. The promise is greater efficiency and consistency in decision-making, reducing the subjective judgments that can lead to unfair outcomes.

But the reality is more complex. The algorithms learn from historical patterns in disability determinations, and these patterns reflect decades of discriminatory decision-making by human adjudicators. Communities of color have historically faced greater barriers to disability benefit approval, often due to disparities in access to medical care, differences in how symptoms are documented and described, and implicit bias in medical and administrative assessment processes.

When algorithmic systems learn from this biased data, they perpetuate discriminatory patterns while claiming mathematical objectivity. The Social Security Administration's algorithms systematically underestimate disability severity in communities of color, directing resources away from populations that have historically faced barriers to benefit access. Automation achieves efficiency, but it does so by encoding discrimination into mathematical formulas that are harder to challenge than human decisions.

Similar patterns emerge across federal benefit programs. The Department of Agriculture's implementation of algorithmic systems for SNAP (food stamp) eligibility determination has systematically disadvantaged rural communities and communities of color, often due to data quality issues and algorithmic assumptions that don't account for regional variations in employment patterns and economic conditions.

The Department of Veterans Affairs has deployed algorithmic risk assessment tools to prioritize mental health services for veterans, but these systems exhibit systematic bias against women veterans and veterans of color, often because the training data reflects historical patterns of underutilization of VA services by these populations. The algorithms interpret lower historical service utilization as indicating lower need, creating self-reinforcing cycles of exclusion.

The Internal Revenue Service's algorithmic audit selection systems demonstrate how administrative efficiency can perpetuate economic inequality. The IRS uses machine learning algorithms to identify tax returns with high likelihood of generating additional revenue through audits, but these systems systematically target low-income taxpayers claiming the Earned Income Tax Credit while under-auditing wealthy taxpayers with complex returns that require more time and expertise to review.

The algorithmic bias occurs because auditing low-income taxpayers is more efficient from the IRS's perspective—their returns are simpler to review and they are less likely to have resources to mount sustained legal challenges to audit findings. The algorithm learns that auditing poor people generates revenue with minimal administrative costs, so it systematically directs enforcement resources toward the populations least able to resist.

International Policy Comparison: Alternative Models of Technology Governance

The contradictions in American technology policy become even more apparent when examined in comparison to approaches adopted by other developed democracies. European Union policies, in particular, offer alternative models that prioritize public interest goals over pure technological capability or economic efficiency.

The EU's General Data Protection Regulation (GDPR), implemented in 2018, represents a fundamentally different approach to technology governance that prioritizes individual privacy rights and democratic oversight over technological innovation or economic competitiveness. Under GDPR, individuals have the right to know when algorithmic systems are making decisions about them, to understand the logic behind those decisions, and to challenge automated determinations that significantly affect their lives.

These requirements create friction in algorithmic decision-making systems, making them more expensive to develop and deploy. American technology companies initially resisted GDPR compliance, arguing that privacy protections would stifle innovation and reduce the competitiveness of European technology industries. But the regulation has demonstrated that robust privacy protections are compatible with technological development—they simply require different approaches that prioritize transparency and accountability over pure efficiency.

The EU's Digital Services Act, which took effect in 2024, extends these principles to platform governance and content moderation systems. Large technology platforms operating in Europe must provide algorithmic transparency, submit to independent audits, and implement risk assessment procedures for their automated systems. The legislation recognizes that algorithmic systems have social and political impacts that extend beyond their immediate economic functions.

Most significantly, the EU's proposed Artificial Intelligence Act represents the world's first comprehensive regulatory framework for AI systems, with provisions that directly address the discriminatory impacts that characterize algorithmic Jim Crow. The legislation bans AI applications that pose unacceptable risks to fundamental rights, requires transparency and human oversight for high-risk AI systems, and establishes enforcement mechanisms that can impose significant penalties for non-compliance.

The AI Act explicitly prohibits government social scoring systems and real-time biometric identification in public spaces, applications that are common in American policing and surveillance programs. It requires algorithmic impact assessments for AI systems used in employment, education, and social services—precisely the areas where American algorithms exhibit systematic bias against communities of color.

European approaches to data center regulation also offer alternatives to the environmental racism that characterizes American infrastructure deployment. Several EU countries have implemented comprehensive environmental impact assessment requirements for data centers, with mandatory consideration of cumulative impacts on already burdened communities. Some jurisdictions require data centers to demonstrate positive local benefits, such as district heating systems that use waste heat from servers to warm nearby buildings.

Nordic countries have developed innovative approaches to sustainable data center development that prioritize environmental protection alongside technological capability. Iceland and Norway have attracted data center investment through abundant renewable energy and cool climates that reduce cooling requirements, but they have also implemented strong environmental protections and community consultation requirements that prevent the environmental racism patterns common in American infrastructure deployment.

The EU Model: Rights-Based Technology Governance

The European Union's approach to technology governance offers a compelling alternative to American policies that prioritize technological capability over social equity. EU regulations are grounded in fundamental rights frameworks that recognize technology as having social and political dimensions that require democratic oversight and accountability.

The GDPR's privacy protections extend beyond individual data rights to encompass collective concerns about algorithmic decision-making and automated profiling. Article 22 of the regulation establishes a general right not to be subject to purely automated decision-making, including profiling, that produces legal effects or similarly significant impacts. This provision directly challenges the administrative automation that characterizes American government algorithms.

Under GDPR, individuals have the right to meaningful information about the logic involved in algorithmic decision-making, not just notification that automated systems are being used. This requirement forces organizations to develop explainable AI systems and to provide transparency about algorithmic operations that are typically treated as trade secrets in American contexts.

The regulation also includes provisions for algorithmic accountability that extend beyond individual rights to collective concerns about discriminatory impact. Data protection authorities can investigate algorithmic systems that exhibit systematic bias, even without individual complaints, and can impose significant penalties for discriminatory automated decision-making.

The Digital Services Act extends these accountability principles to platform governance and content moderation systems. Large platforms must conduct risk assessments that evaluate their impact on fundamental rights, democratic processes, and social cohesion. They must implement mitigation measures for identified risks and submit to independent audits of their algorithmic systems.

These requirements create incentives for technology companies to develop more equitable algorithmic systems, not because of market forces or voluntary corporate responsibility commitments, but because of legal requirements backed by significant enforcement penalties. The approach recognizes that algorithmic systems are not neutral tools but social and political infrastructure that requires democratic governance.

The proposed AI Act represents the most comprehensive attempt to govern artificial intelligence systems according to fundamental rights principles. The legislation categorizes AI applications according to their risk to fundamental rights and social welfare, with prohibited uses, high-risk applications requiring extensive oversight, and general-purpose AI systems subject to transparency requirements.

The prohibited uses category includes applications that are common in American policing and surveillance programs, such as social scoring systems and real-time biometric identification in public spaces. The legislation recognizes that some technological capabilities are incompatible with democratic governance and human rights protection.

High-risk AI systems, including those used in employment, education, law enforcement, and social services, must meet extensive requirements for data quality, transparency, human oversight, and accuracy. These requirements directly address the discriminatory impacts documented in American algorithmic systems, but they do so through legally enforceable mandates rather than voluntary corporate commitments.

Case Study: Political Networks and Infrastructure Extraction

When former government officials move into private consulting, they carry with them something invaluable: inside knowledge of how the system actually works. This revolving door between public service and private profit creates networks that consistently harm the same communities, over and over again.

Consider what happens when someone who spent years managing federal technology contracts suddenly becomes a consultant helping companies navigate those exact same processes. They know which regulators ask hard questions and which ones rubber-stamp applications. They understand how to frame environmental impact assessments to minimize scrutiny. Most importantly, they've learned to identify communities that lack the political clout to fight back effectively.

These career transitions happen quietly, without fanfare or public attention. A communications director here, a policy analyst there—each building expertise that becomes commercially valuable once they leave government. Their new employers aren't necessarily seeking corruption; they want efficiency. And efficiency, in this context, means finding the path of least resistance.

The communities that bear the brunt of this "efficient" decision-making share predictable characteristics. They're predominantly Black and Brown. They have high poverty rates and limited political representation. Many have already been sacrificed for industrial development in previous generations, making them seem like natural choices for the next wave of environmental burden.

What makes this pattern particularly insidious is its apparent neutrality. Nobody explicitly says "target the Black community." Instead, consulting firms develop sophisticated matrices that evaluate "community acceptance factors" and "regulatory complexity indicators." These euphemistic frameworks consistently point toward the same types of neighborhoods, creating plausible deniability for decisions that perpetuate racial injustice.

The xAI facility in South Memphis exemplifies this process. The site selection appeared to follow standard business criteria—available land, existing industrial zoning, proximity to electrical infrastructure. Yet these "neutral" factors inevitably led to a historically Black community already burdened with cancer rates four times the national average.

Meanwhile, when communities have organized successfully against unwanted development, companies often withdraw entirely rather than modify their plans. This pattern of "reverse redlining" reveals the calculated nature of site selection: corporations pursue projects only where resistance seems manageable.

The human cost of these decisions extends far beyond abstract policy discussions. Families watch their neighborhoods become dumping grounds for society's most polluting industries. Children develop asthma from increased truck traffic and industrial emissions. Elderly residents see property values collapse as their communities become associated with environmental hazards.

These consulting networks operate largely in shadow, through handshake agreements and informal relationships rather than registered lobbying contracts. Their influence becomes visible only through the consistent patterns of harm they facilitate—patterns that span multiple administrations and persist regardless of which party controls government.

The tragedy lies not just in individual cases of environmental injustice, but in how these networks institutionalize racism through seemingly technical processes. They transform political rhetoric about American technological leadership into a system that extracts resources from the most vulnerable communities while concentrating benefits among the already powerful.

Breaking these patterns requires more than policy reform; it demands fundamental changes in how we think about community consent and environmental justice in an age of expanding technological infrastructure.

The Failure of Technological Nationalism

The contradiction between America First rhetoric and infrastructure dependency reflects a broader failure of technological nationalism as an approach to contemporary policy challenges. Technological capabilities cannot be separated from the social and environmental contexts in which they are developed and deployed, and attempts to achieve technological dominance through purely national approaches inevitably reproduce patterns of inequality and exploitation.

American technology policy operates under the assumption that technological superiority can be achieved through market mechanisms and private sector innovation, with government playing a supporting role through procurement, subsidies, and regulatory frameworks that favor domestic companies. This approach treats technology as a national asset that can be developed and controlled through appropriate policy incentives.

But this nationalist framework misunderstands how contemporary technology systems actually operate. AI capabilities depend on global supply chains, international data flows, and infrastructure networks that transcend national boundaries. Even when technology companies are nominally American, their operations depend on resources, labor, and infrastructure that are distributed globally.

More fundamentally, technological capabilities cannot be separated from the social contexts in which they are developed and deployed. American AI systems exhibit systematic bias against communities of color not because of technical failures, but because they are trained on data that reflects centuries of discriminatory policies and social practices. These biases cannot be addressed through technical fixes or better algorithms—they require confronting the underlying social inequalities that the technology systems help perpetuate.

The environmental racism that characterizes American technology infrastructure deployment is not an unfortunate side effect of pursuing technological capabilities, but a systematic strategy for minimizing political resistance to environmentally burdensome facilities. Technology companies locate their most polluting operations in communities of color because these communities have less political power to resist, not because of any technical requirements.

Technological nationalism fails because it treats these social and environmental impacts as external to technology development rather than recognizing them as integral features of how technological systems operate. American AI capabilities depend on environmental exploitation of marginalized communities, and attempts to maintain technological leadership while ignoring these impacts ultimately undermine both technological development and social cohesion.

The alternative is not to abandon technological development, but to recognize that genuine technological capabilities require addressing the social and environmental contexts in which technology operates. This means developing AI systems that do not exhibit systematic bias, deploying infrastructure in ways that benefit rather than burden local communities, and recognizing that technological development is a social process that requires democratic governance and accountability.

Beyond "America First": Toward Democratic Technology Governance

The contradictions in American technology policy point toward alternative approaches that prioritize democratic governance and social equity alongside technological capability. These alternatives are not utopian visions but practical policies that other democratic countries have already begun implementing.

Democratic technology governance begins with recognizing that algorithmic systems are not neutral tools but social and political infrastructure that requires public oversight and accountability. This means moving beyond market-based approaches that treat technology companies as private actors toward regulatory frameworks that recognize the public character of algorithmic decision-making.

The European Union's approach to AI regulation offers one model for democratic technology governance, with legally enforceable requirements for transparency, accountability, and non-discrimination in algorithmic systems. But effective technology governance also requires addressing the infrastructure dependencies that enable environmental racism and community exploitation.

This means developing public alternatives to private technology infrastructure, including government-operated data centers that serve public purposes rather than corporate profits. Several countries have begun experimenting with public cloud computing services that prioritize privacy protection and democratic accountability over economic efficiency or technological capabilities.

Public infrastructure development also creates opportunities to address environmental justice concerns through strategic siting decisions that benefit rather than burden local communities. Data centers can be designed to provide district heating, renewable energy generation, and high-paying employment opportunities for local residents rather than serving as sources of environmental burden.

Community ownership models offer another alternative to corporate-controlled technology infrastructure. Cooperative data centers, municipal broadband networks, and community-controlled digital platforms demonstrate that technological capabilities can be developed through democratic processes that prioritize local needs and benefits.

These alternatives require significant public investment and political commitment, but they offer pathways toward technological development that serves democratic purposes rather than corporate profits. They also provide models for international co-operation based on shared democratic values rather than competitive technological nationalism.

The choice is not between technological capability and social equity, but between technology systems that serve democratic purposes and those that perpetuate inequality and exploitation. American technology policy has chosen the latter path, but the contradictions this creates offer opportunities for developing more equitable alternatives.

The future of American technology depends not on achieving dominance through competitive strategies that exploit marginalized communities, but on developing technological capabilities that serve democratic purposes and contribute to social equity. This requires abandoning the nationalist frameworks that justify environmental racism and infrastructure extraction in favor of approaches that recognize technology as social infrastructure requiring democratic governance.

The DOGE-xAI partnership represents the logical conclusion of technological nationalism—a government that achieves computational capabilities by partnering with corporations whose operations depend on environmental exploitation of the communities that government is supposed to serve. The alternative is technology policy that serves democratic purposes but achieving this requires confronting the power structures that the current system is designed to perpetuate.

As the environmental and social costs of current technology policies become increasingly visible, the contradictions between America First rhetoric and infrastructure dependency will become harder to ignore. The question is whether American democracy has the capacity to develop alternatives, or whether technological nationalism will continue to reproduce the patterns of inequality and exploitation that algorithmic Jim Crow was designed to address.

CHAPTER 8: PLATFORM GOVERNANCE & CORPORATE RESPONS

In September 2020, more than 200 personal accounts, many of which were administrators of environmental and Indigenous groups and pages, found themselves suspended by Facebook for several days. The suspensions occurred before a planned protest targeting private equity company KKR & Co., with Indigenous organizers locked out of their Facebook event pages, unable to post or send messages (Folley, 2020). These accounts, which activists told *The Guardian* were suspended for an alleged "intellectual property rights violation," represented a pattern of systematic suppression targeting environmental justice content. In two 2019 computational linguistic studies, researchers discovered that AI intended to identify hate speech may end up amplifying racial bias. In one study, researchers found that tweets written in African American English commonly spoken by Black Americans are up to twice more likely to be flagged as offensive compared to others. Global Witness, an organization tracking digital threats against activists, found that Facebook was cited by 62% of over 200 environmental and land rights defenders as the platform where they experienced online abuse, followed by X (formerly Twitter) at 37%, WhatsApp at 36%, and Instagram at 26% (Global Witness, 2024). This systematic targeting represents a broader pattern where platform governance suppresses environmental justice content while simultaneously subjecting affected communities to environmental burdens from platform infrastructure.

Platform governance operates as a sophisticated system of digital segregation through interconnected mechanisms that mirror historical patterns of exclusion. Barocas and Selbst define online proxies as "factors used in the scoring process of an algorithm which are mere stand-ins for protected groups, such as zip code as proxies for race, or height and weight as proxies for gender". These systems concentrate environmental costs in marginalized communities while providing superior digital services to affluent areas, creating what might be termed "algorithmic Jim Crow" operating through corporate infrastructure decisions.

Platform Infrastructure & Environmental Burden Distribution

Platform companies exercise control over information flow through infrastructure systems that follow predictable patterns of environmental racism. Data centers moving into predominantly communities of color and low-income residential areas can exacerbate historical trends that overlook community health for industrial needs; choosing locations as sacrifice zones, creating public harm for private gain that disproportionately affects already vulnerable populations (Marrinan, 2024).

Large data centers can consume up to 5 million gallons per day, equivalent to the water use of a town populated by 10,000 to 50,000 people. These facilities operate 24/7, generating noise pollution that disrupts sleep patterns and air pollution that exacerbates existing health disparities in host communities (Environmental and Energy Study Institute, n.d.). For communities like Brenda's and David's, the computational whir of data centers is not merely an annoyance, but a source of mental and physical harm. Brenda, a nurse by training, reported an uptick in her blood pressure and cortisol levels with the onset of the noise. (Monserrate, 2022).

The environmental costs prove staggering while communities receive inferior platform services. Data centers and data transmission networks are responsible for nearly 1% of energy-related greenhouse gas (GHG) emissions, which contribute to rising global temperatures and climate change. Yet these same communities experience slower content loading, reduced algorithmic visibility, and degraded digital services—a pattern reproducing historical segregation through technological means.

Platform infrastructure placement decisions reflect strategic corporate calculations about political resistance capacity. Companies locate environmentally burdensome facilities in communities with limited political power while ensuring optimal service delivery for affluent users. Black, low-income Americans already have the highest mortality rate from exposure to PM2.5, a type of air pollution produced by electricity generation. Platform infrastructure compounds these existing environmental health disparities (Marrinan, 2024).

Algorithmic Enforcement & Differential Treatment

Content moderation algorithms function as sophisticated digital policing systems producing discriminatory outcomes at unprecedented scale. Messages containing minority identity markers (e.g., Arabs, Black, LGBT+), messages with explicit swearwords, and messages with specific racial and ethnic dialects (e.g., Black-aligned English) have a higher probability to be misclassified as toxic regardless of their communicative contexts (Ng et al., 2024).

Instagram influencers of marginalized identities and subjectivities, for example those that are plus-sized or POC, often express through their social media that their content is moderated more heavily and will sometimes place blame on what they call "the algorithm" as the source of their feelings of discrimination. These experiences reflect systematic rather than incidental discrimination (Albury, 2024).

Shadow-bans are "a form of online censorship where you're still allowed to speak, but hardly anyone gets to hear you," *The Washington Post* explained. Their content might not be removed, but some creators notice that engagement with their posts plummets outside of their immediate friends. This algorithmic suppression creates economic consequences for communities already facing systematic exclusion (Coleman, 2023).

The enforcement patterns mirror discriminatory policing practices. Many Black Lives Matter (BLM) activists were similarly frustrated when Facebook flagged their accounts, but didn't do enough to stop racism and hate speech against Black people on their platform. Platform algorithms determine which businesses receive visibility, which creators earn revenue, and which communities access economic opportunities—decisions that concentrate benefits among already-privileged groups while limiting opportunities for communities of color.

A study published by the Anti-Defamation League in 2025 found that several major Large language models (LLMs), including ChatGPT, Llama, Claude, and Gemini showed antisemitic bias. These biases operate across content types and platforms, creating systematic barriers to economic mobility and democratic participation for marginalized communities.

Information Ecosystem Fragmentation & Digital Redlining

Platform recommendation algorithms create distinct information environments that separate users along racial and economic lines. Machine learning moderation compares content with existing data, which means unique content needs to be already normative, or at least 'kncwn' for machine learning moderation to 'see' it as a constitutive element to prompt action, such as deletion (Albury, 2024).

This algorithmic segregation affects access to essential information during public health crises, economic opportunities, and political participation. When officers in overpoliced neighborhoods record new offer ces, a feedback loop is created, whereby the algorithm generates increasingly biased predictions targeting these neighborhoods. In short, bias from the past leads to bias in the future. (UN Office of the High Commissioner for Human Rights, 2024).

Geographic targeting algorithms assume users in majority-minority ZIP codes prefer different content types, creating distinct information flows based on residential patterns shaped by historical redlining. Facebook s advertising platform contained proxies that allowed housing marketers to micro-target preferred renters and buyers by clicking off data points, including zip code preferences. These systems reproduce housing discrimination in digital spaces while limiting access to economic opportunities.

The information segregation operates internationally, with platform algorithms treating users in Global South countries as sources of data extraction rather than information consumers. Communities generate data that improves platform services for wealthy users while receiving inferior algorithmic treatment themselves—a pattern reflecting broader digital colonialism.

Corporate Power & Accountability Gaps

Platform governance operates through corporate structures that systematically exclude affected communities from infrastructure and algorithmic decisions. Unlike traditional public utilities, platform companies face minimal oversight regarding environmental impact, service equity, or community consultation. Facebook recently completed a civil rights audit to determine its handling of issues and individuals from protected groups. After the reveal of how the platform was handling a variety of issues, including voter suppression, content moderation, privacy, and diversity, the company has committed to an updated audit around its internal infrastructure to handle civil rights grievances (Brookings Institution, 2023).

Corporate responsibility frameworks focus primarily on shareholder value and regulatory compliance rather than community impact assessment or environmental justice considerations. This governance gap enables platforms to externalize environmental and social costs onto marginalized communities while concentrating economic benefits in affluent areas. "Bias allegations against social media platforms have rarely succeeded in court," *The Verge* noted. YouTube won two lawsuits from LGBTQ+ and Black video creators who alleged algorithmic discrimination. (Coleman, 2023).

The result creates sophisticated discrimination systems operating with plausible deniability through technical complexity and corporate responsibility rhetoric. Community members face algorithmic bias, environmental burdens, and economic exclusion while lacking meaningful recourse through existing institutional channels.

International Regulatory Responses & Their Limitations

Comparative analysis reveals alternative approaches that better protect community interests, though implementation faces significant obstacles. The act requires all intermediaries to publish annual transparency reports on content restrictions, government requests for user data, and the use of automated moderation tools. The DSA requires platforms to put in place measures to counter the spreading of illegal goods, services or content online, such as mechanisms for users to flag such content and for platforms to cooperate with "trusted flaggers." (Freedom House, 2024; European Commission, n.d.).

The DSA aims to end an era in which tech companies have essentially regulated themselves – setting their own policies on how to moderate content, and issuing "transparency reports" about their efforts to combat harms like disinformation that have been practically impossible for third parties to scrutinize. The Act requires platforms to conduct risk assessments and provide community oversight mechanisms for content moderation decisions (Algorithm Watch, n.d.).

Key requirements include disclosing to regulators how their algorithms work, providing users with explanations for content moderation decisions, and implementing stricter controls on targeted advertising. It also imposes specific rules on "very large" online platforms and search engines (those having more than 45 million monthly active users in the EU). These regulations demonstrate feasible alternatives to current American approaches while revealing political obstacles preventing similar reforms.

However, The European Commission intensified its investigation into X's content moderation practices earlier this year by demanding internal documents related to the platform's algorithms. Specifically, the EU's executive arm requested that X provide documentation about its recommender system and any recent modifications made to it. Enforcement challenges remain significant, particularly regarding international platform operations (Jahangir, 2025).

American platform companies lobbied successfully against environmental review requirements, community consultation mandates, and algorithmic transparency regulations that operate in other countries. The selective regulatory attention reflects broader failures in technology governance that enable corporate exploitation while maintaining accountability gaps.

Community Resistance & Environmental Justice Integration

Communities resist platform discrimination through organizing strategies combining technical analysis with traditional community advocacy. Through analyzing the dynamics between the #IWantToSeeNyome campaign (The #IWantToSeeNyome campaign represents one of the most successful examples of community organizing against algorithmic discrimination in recent years.) and the reactions from Instagram's head Adam Mosseri, it becomes clear how this process affects certain user's perceptions of content moderation as racially biased and fatphobic. These efforts require sustained community engagement and technical skill development (Albury, 2024).

To avoid the looming threat of shadow-banning, some content creators have taken to using workarounds "such as not using certain images, keywords or hashtags or by using a coded language known as algospeak," *The Post* explained. However, individual workarounds cannot address systematic discrimination requiring collective organizing and policy advocacy (Coleman, 2023).

Platform governance reform cannot succeed independently from broader environmental justice efforts. AI infrastructure may exacerbate past harms that previous fossil fuel industries have imposed upon communities of color, excusing its extractive and harmful impacts in the name of progress and innovation. The same communities facing discriminatory algorithmic treatment also bear disproportionate environmental burdens from platform infrastructure (Marrinan, 2024).

Communities near proposed or current data centers often object to their construction or operation due to concerns about noise, greenhouse gas emissions, strain on local utilities, utility costs, loss of cultural areas, and air quality issues (Gradient Corporation, 2024). Reform efforts must address these connections rather than treating platform policies separately from environmental concerns.

Policy Integration & Systemic Reform

Platform governance reform requires integration with broader efforts addressing algorithmic discrimination and environmental racism. "Artificial intelligence technology should be grounded in international human rights law standards," Dr. Ashwini K.P. said. "The most comprehensive prohibition of racial discrimination can be found in the International Convention on the Elimination of All Forms of Racial Discrimination." (UN Office of the High Commissioner for Human Rights, 2024).

Current regulatory approaches fail because they fragment platform governance into separate issues: content policy, data privacy, antitrust enforcement, and environmental regulation. This fragmentation enables platform companies to address individual concerns while maintaining overall systems of exploitation and discrimination.

By continually improving algorithmic review and regulatory processes, we can better address the algorithmic challenges of modern society through comprehensive frameworks addressing platform operations holistically (Wang et al., 2024). Environmental impact assessments should include algorithmic bias analysis, while civil rights enforcement should incorporate environmental justice concerns.

International coordination proves essential because platform operations cross national boundaries enabling companies to evade accountability by relocating operations to countries with permissive regulatory environments. Reform efforts must address corporate structures prioritizing shareholder returns over community wellbeing through public ownership options, cooperative governance structures, and regulatory frameworks requiring community benefit agreements.

Platform Justice as Infrastructure Justice

Platform governance represents a critical component of broader struggles for infrastructure justice and community self-determination. The infrastructure decisions, algorithmic systems, and corporate structures governing major digital platforms create systematic discrimination intersecting with housing segregation, environmental racism, and economic exclusion documented throughout this analysis.

Women, Black people, and Jews are often targets of online hate speech. Online hate proliferates where human content moderators miss offensive content. Also, algorithms are prone to errors. They may multiply errors over time and may even end up promoting online hate. Yet communities resist these systems through organizing efforts combining technical analysis with traditional community advocacy, demonstrating possibilities for platform governance serving community needs (European Union Agency for Fundamental Rights, 2024).

The alternatives exist at every level: community-controlled infrastructure, democratic platform governance, regulatory frameworks prioritizing environmental and social justice, and international cooperation based on mutual aid rather than corporate extraction. Implementation requires sustained community organizing, policy advocacy, and resource investment challenging corporate power while building community capacity for technological self-determination.

Without such transformation, digital platforms will continue operating as sophisticated systems for concentrating wealth and power while externalizing costs onto marginalized communities. Platform governance reform represents both practical necessity for community survival and strategic opportunity for building broader movements toward economic and environmental justice.

The question is not whether alternatives are possible—communities worldwide demonstrate their feasibility through ongoing resistance efforts. The question is whether American communities can build sufficient political power to overcome corporate resistance and implement platform governance systems prioritizing human dignity over shareholder profits. That answer depends on continued community organizing, sustained policy advocacy, and international cooperation recognizing platform justice as inseparable from broader struggles for community self-determination and environmental justice.

CHAPTER 9: EDUCATIONAL TECHNOLOGY & ACADEMIC SURVE

E ducational technology promises democratization—computational systems that expand access, reduce bias, and create equitable learning environments. Universities market algorithmic admissions as tools identifying overlooked talent while eliminating human prejudice. School districts promote surveillance technologies as safety measures enhancing learning outcomes. Assessment platforms claim objective evaluation transcending cultural barriers.

Yet beneath these narratives lies a troubling reality: educational algorithms have become sophisticated instruments perpetuating the inequalities they claim to address. Rather than expanding opportunity, these systems function as digital gatekeepers systematically excluding students from marginalized communities while providing institutional cover for discriminatory practices.

This chapter examines how algorithmic systems within educational institutions reproduce exclusion patterns mirroring historical academic discrimination. From graduate admissions disadvantaging underrepresented applicants, to surveillance technologies criminalizing normal student behavior in schools serving predominantly Black and Hispanic populations, to assessment tools interpreting cultural differences as academic deficiencies, educational algorithms create new mechanisms for digital redlining in academic settings.

This technological transformation spans all education levels. Elementary schools deploy behavioral monitoring systems tracking student movements and flagging "concerning" activities disproportionately affecting students of color. High schools implement predictive analytics channeling students into academic tracks based on algorithmic assessments of "potential." Universities rely on automated screening filtering thousands of applications before human review, often eliminating qualified applicants whose backgrounds don't match algorithmic definitions of academic promise.

These systems operate beneath mathematical objectivity's veneer. When human administrators make discriminatory decisions, such actions can be identified and challenged through accountability mechanisms. When algorithms produce identical outcomes, institutions attribute results to neutral computational analysis rather than biased judgment, creating "algorithmic laundering" of discriminatory practices.

Students most affected often lack technical knowledge or institutional access to understand how algorithmic processes shape educational opportunities. They experience rejection, surveillance, or tracking without comprehending the computational logic governing these decisions. Meanwhile, algorithms continue operating through proprietary methodologies protected by trade secrets, making external audit nearly impossible.

This creates a profound paradox: universities producing groundbreaking algorithmic bias research simultaneously deploy biased systems controlling program access. Scholars exposing how machine learning perpetuates discrimination often navigated these systems to reach positions where such research becomes possible. Their critical work emerges from institutions employing discriminatory algorithms to limit their initial access.

The implications extend beyond individual exclusion to shape knowledge production itself. When algorithmic gatekeepers systematically exclude researchers from marginalized communities, they limit diverse perspectives contributing to scholarship on technology, bias, and social justice. This creates feedback loops where communities most affected by algorithmic discrimination have least representation among those studying these systems.

Educational algorithms must be understood not as neutral tools but as systems encoding existing power structures through computational processes. These technologies transform social hierarchies into algorithmic logic, perpetuating inequality while claiming scientific legitimacy. Mathematical frameworks underlying educational algorithms often reflect biases in historical data, institutional practices, and design choices by developers lacking awareness of discriminatory impacts.

Educational surveillance technologies add another dimension. Schools increasingly monitor student behavior through digital platforms tracking online activity to physical movements. These systems generate massive datasets feeding algorithmic assessments of academic potential and future success. Yet surveillance technologies deploy unevenly, with schools serving predominantly Black and Hispanic students experiencing higher monitoring and punitive intervention levels.

This chapter traces these patterns across educational contexts, examining how algorithmic systems create new academic surveillance forms while perpetuating historical exclusion patterns. Through analysis of admissions algorithms, assessment technologies, and surveillance systems, we reveal how educational technology functions not as democratizing force but as digital discrimination mechanism operating through computational neutrality claims.

The goal extends beyond documenting harmful practices to understanding connections with broader technological discrimination patterns affecting marginalized communities. Educational algorithms represent a critical component of "algorithmic Jim Crow"—systematic use of computational systems perpetuating racial inequality while maintaining plausible deniability about discriminatory intent.

By examining these systems within educational institutions, we better understand mechanisms through which algorithmic bias reproduces across generations, limiting opportunities for students who might contribute critical perspectives on technology, justice, and social change. Only through such analysis can we develop resistance and transformation strategies centering community needs over technological efficiency.

The Doctoral Pipeline Problem

The author's research on African American representation in doctoral programs reveals "Pretty huge Discrepancies" (Ph.D) that expose how algorithmic systems create systematic barriers to academic advancement throughout the educational pipeline. The 2023 data on doctoral recipients reveals a troubling pattern: while African Americans represent approximately 13% of the U.S. population, their representation among doctoral recipients falls dramatically short in most fields, with particularly stark underrepresentation in STEM disciplines.

The data reveals that computer and information sciences shows one of the smallest proportions of African American PhD recipients among all fields, with what the National Science Foundation data characterizes as a "very small yellow bar" representing Black doctoral recipients compared to substantially larger representation in biological sciences, engineering, or psychology. This severe underrepresentation in computer science—the field increasingly central to algorithmic decision-making—means that the very people most affected by algorithmic bias are systematically excluded from the research communities developing these systems.

This isn't a "leaky pipeline," but a systematically designed filtering system that uses algorithmic processes appearing neutral while producing discriminatory outcomes. The concentration of African American doctoral recipients in non-Science and Engineering fields, while showing "stronger representation in humanities, education, and related fields," suggests tracking patterns that begin early in educational pathways and channel students of color away from the STEM disciplines that increasingly govern social and economic opportunities.

The filtering begins early, but compounds dramatically as students progress through educational systems that increasingly rely on algorithmic decision-making. Educational data mining systems analyze student performance data to predict academic success and make recommendations about course placement, but these systems systematically underestimate students from communities historically excluded from academic opportunities. When algorithms learn from decades of biased educational data, they reproduce patterns where students of color are consistently steered away from advanced mathematics and science courses that serve as prerequisites for competitive STEM programs.

The author's analysis reveals that the barriers become most visible in undergraduate research opportunities, which are crucial pathways to graduate school but are increasingly allocated through algorithmic systems exhibiting systematic bias. Research experience programs use automated screening to evaluate applications based on GPA thresholds and standardized test scores, but these metrics systematically exclude students whose academic records don't conform to traditional patterns of academic success despite demonstrating strong research interest and capability.

The National Science Foundation data analyzed by the author shows that fields with the lowest African American representation—computer and information sciences, mathematics and statistics, geosciences, and agricultural sciences—are precisely those where algorithmic decision-making is becoming most prevalent. This creates a feedback loop where the communities most affected by algorithmic bias are systematically excluded from the academic disciplines that could address these problems.

Research consistently documents the persistent underrepresentation of African Americans among doctoral recipients in STEM fields, revealing systemic barriers including limited early educational opportunities, insufficient mentorship, financial constraints, and institutional biases. However, the problem extends beyond individual barriers to encompass what can be understood as "epistemic exclusion"—the systematic exclusion of African American perspectives from knowledge production itself.

When African American students represent such small numbers in doctoral programs, particularly in computer science and mathematics, entire domains of inquiry remain unexplored and methodological innovations go undeveloped. The absence of African American scholars from fields developing algorithmic systems means that the perspectives of communities most harmed by these technologies are systematically excluded from their design and governance. This creates what the author's research identifies as "significant blind spots in disciplinary knowledge and diminished attention to issues affecting Black communities."

The rollback of Diversity, Equity, and Inclusion (DEI) initiatives threatens to exacerbate these disparities by dismantling support structures designed to counteract historical exclusion. As the author's research warns, "these retrenchments—often framed as responses to political pressure—risk sacrificing the intellectual vibrancy, innovation, and basic fairness that diverse doctoral cohorts bring to knowledge production and academic communities." The timing is particularly troubling as algorithmic systems become more pervasive and consequential for communities of color.

This creates a self-fulfilling prophecy where algorithmic systems reduce opportunities for students of color to develop research experience and relationships with faculty mentors, making them less competitive for graduate programs, which reinforces the algorithmic prediction that such students are poor candidates for doctoral study.

The impact becomes most visible in undergraduate research opportunities, which are crucial pathways to graduate school but are increasingly allocated through algorithmic systems that exhibit systematic bias. Research experience programs use automated screening to evaluate applications based on GPA thresholds, standardized test scores, and algorithmic assessments of research potential.

These screening systems systematically exclude students whose academic records don't conform to traditional patterns of academic success, even when those students demonstrate strong research interest and capability. Students who work part-time jobs to support their families may have lower GPAs despite strong intellectual abilities, while students from schools with limited Advanced Placement offerings may appear less competitive despite equivalent academic preparation.

The algorithmic screening creates barriers that are particularly severe for first-generation college students and students from underrepresented minorities, who are less likely to understand the importance of research experience for graduate school preparation and less likely to have family networks that can provide guidance about navigating academic systems.

The author's analysis of National Science Foundation data on doctoral program participation reveals the cumulative impact of these algorithmic barriers. African American students who successfully navigate the undergraduate screening processes and gain admission to doctoral programs often excel academically, contradicting algorithmic predictions about their likelihood of success. But their numbers remain small because of the systematic filtering that occurs at earlier stages of the academic pipeline.

The problem is not that African American students lack academic capability or interest in doctoral study, but that algorithmic systems create barriers to accessing the preparatory experiences and opportunities that enable competitive graduate school applications. The algorithms perpetuate a cycle where underrepresentation becomes evidence for continued exclusion.

Algorithmic Admissions: The New Gatekeepers of Higher Education

The digitization of American education has created unprecedented opportunities for algorithmic discrimination to operate at institutional scale. Educational technology systems—from admissions algorithms to student assessment platforms—have become sophisticated mechanisms for perpetuating racial inequities while maintaining the appearance of objective, merit-based evaluation. These systems represent a particularly insidious form of digital discrimination because they operate at the gateway to economic mobility, determining which students gain access to educational opportunities that could disrupt cycles of inequality.

The Architecture of Algorithmic Gatekeeping

Educational institutions increasingly deploy algorithmic systems as gatekeepers, making initial screening decisions about student applications, academic progress, and resource allocation. These systems promise efficiency and objectivity in managing massive volumes of applications and assessments, but their operational logic embeds historical patterns of exclusion into seemingly neutral technical processes.

The transformation occurs through what can be understood as *algorithmic credential laundering*—the process by which discriminatory outcomes are legitimized through technical complexity and claims of objectivity. When a human admissions officer rejects a qualified applicant from an underrepresented background, the decision appears subjective and potentially discriminatory. When an algorithm makes the same decision based on "objective" metrics, the outcome appears scientifically justified.

This technical legitimation obscures how algorithmic systems systematically disadvantage students from marginalized communities by incorporating proxy variables that correlate with race and socioeconomic status. The algorithms learn from historical data that reflects decades of discriminatory practices, then reproduce those patterns with mechanical precision while claiming to eliminate human bias.

Algorithmic Admissions: Mechanizing Exclusion

University admissions represent one of the most consequential applications of algorithmic decision-making in education. Major research universities process tens of thousands of applications annually, creating economic pressure to adopt automated screening systems that can efficiently filter candidates before human review. These systems analyze standardized test scores, grade point averages, course selections, and other quantitative measures to generate composite rankings that predict academic success.

The apparent objectivity masks systematic biases embedded in the data and metrics these systems use. Historical admissions data reflects institutional practices that have systematically excluded students from underrepresented communities. When algorithms learn from this data, they perpetuate discriminatory patterns while appearing to eliminate human prejudice.

The University of Texas at Austin's experience illustrates these dynamics. The institution implemented automated screening systems designed to identify top candidates from massive applicant pools. The algorithm analyzed high school GPAs, standardized test scores, class rank, and other academic indicators to generate rankings for initial screening decisions.

Analysis of the system's outcomes revealed systematic disparities in how the algorithm evaluated students from different backgrounds. Students from well-resourced suburban high schools consistently received higher algorithmic rankings than students with comparable academic records from urban schools serving predominantly students of color. The disparity occurred because the algorithm incorporated contextual factors like high school performance indicators and course availability that systematically disadvantaged students from under-resourced schools.

Students who earned exceptional GPAs in schools with limited Advanced Placement offerings were ranked lower than students with equivalent GPAs from schools with extensive AP programs. This algorithmic logic penalized students for attending under-resourced schools, ignoring research indicating that students who excel in challenging environments often demonstrate greater academic resilience and potential for success.

Graduate School Gatekeeping & Doctoral Diversity

Graduate school admissions exhibit particularly severe patterns of algorithmic discrimination, with devastating impacts on doctoral program diversity. The Graduate Record Examination (GRE), widely used in algorithmic screening for graduate programs, exhibits systematic racial and socioeconomic disparities that correlate with historical exclusion rather than academic potential or research capability.

Research demonstrates that standardized tests like the GRE exhibit persistent score gaps that correlate with historical patterns of educational exclusion rather than academic potential.

When graduate programs use algorithmic systems that heavily weight GRE scores, they systematically exclude qualified candidates from underrepresented communities. Students who demonstrate strong research capabilities, intellectual curiosity, and academic achievement through other measures are filtered out by algorithms that prioritize standardized test performance over more meaningful indicators of graduate school potential.

The bias proves particularly problematic in STEM fields, where doctoral program diversity remains crucial for addressing historical underrepresentation and developing research that serves diverse communities. Algorithmic admissions systems in engineering, computer science, and physical sciences programs systematically reduce admission rates for underrepresented minorities, perpetuating demographic homogeneity that limits these fields' capacity to address societal challenges.

Analysis of STEM doctoral programs reveals declining representation of underrepresented minorities following implementation of algorithmic screening systems that heavily weight standardized test scores.

Research indicates that GRE scores correlate poorly with success in doctoral programs, particularly for students from underrepresented backgrounds who may have limited access to test preparation resources but possess strong research potential. Despite this evidence, many programs continue using algorithmic systems that prioritize GRE performance, suggesting that these systems serve to maintain exclusionary practices rather than improve admissions quality.

Professional School Discrimination

Medical school admissions demonstrate similar patterns of algorithmic bias, with automated screening systems that systematically disadvantage students from backgrounds underrepresented in medicine. The Medical College Admission Test (MCAT) exhibits racial and socioeconomic disparities comparable to other standardized assessments, and when medical schools rely heavily on algorithmic screening based on MCAT scores, they reduce admission rates for students from communities underserved by the medical profession.

The Medical College Admission Test (MCAT) exhibits similar patterns of racial and socioeconomic disparities documented across standardized assessments, with systematic score gaps that persist despite controlling for other academic indicators.

The impact extends beyond individual admission decisions to shape the demographic composition of entire professional fields. When algorithmic systems systematically exclude students from underrepresented communities from medical school, they perpetuate physician workforce disparities that contribute to health inequities in communities of color. Research consistently demonstrates that physicians from underrepresented backgrounds are more likely to practice in underserved communities and provide culturally competent care to patients from similar backgrounds.

Evidence consistently demonstrates that physicians from underrepresented backgrounds are more likely to practice in underserved communities and provide culturally responsive care to patients from similar backgrounds, suggesting that admissions biases perpetuate health disparities.

Law school admissions exhibit parallel patterns, with algorithmic systems relying heavily on Law School Admission Test (LSAT) scores systematically reducing admission rates for underrepresented minorities. The resulting lack of diversity in the legal profession has consequences for equal justice, as communities of color have limited access to attorneys who understand their experiences and can effectively advocate for their interests.

Surveillance Infrastructure in Educational Settings

Beyond admissions, educational institutions increasingly deploy surveillance technologies that disproportionately monitor and discipline students of color. These systems range from facial recognition in school buildings to algorithmic analysis of student behavior patterns, creating digital panopticons that criminalize normal adolescent behavior when exhibited by students from marginalized communities.

School-based surveillance systems exhibit the same environmental racism patterns observed in broader surveillance infrastructure deployment. Schools serving predominantly students of color are more likely to implement comprehensive surveillance systems, including facial recognition cameras, metal detectors, and algorithmic behavior monitoring software. These technologies create prison-like environments that normalize surveillance and control for young people who are disproportionately likely to encounter similar systems throughout their lives.

Educational institutions serving predominantly students of color deploy comprehensive surveillance systems at significantly higher rates than schools in affluent, majority-white districts, creating disparate enforcement environments that normalize surveillance for young people from marginalized communities.

The algorithms used to analyze student behavior exhibit racial bias similar to other automated decision-making systems. Behavior that is interpreted as normal when exhibited by white students may be flagged as threatening or disruptive when exhibited by students of color. These algorithmic interpretations feed into disciplinary systems that push students from marginalized communities out of educational settings and into the juvenile justice system.

Academic Assessment & Algorithmic Bias

Standardized testing systems increasingly rely on algorithmic scoring and analysis that embed racial and cultural biases into assessments of academic achievement. Computer-based testing platforms use natural language processing to evaluate student writing, but these systems exhibit systematic biases against linguistic patterns associated with students from diverse cultural backgrounds.

Research conducted by Educational Testing Service (ETS) on their widely-used E-rater automated scoring engine demonstrates the pervasive nature of algorithmic bias in educational assessment. In studies spanning from 1999 to 2018, ETS consistently found that their system gave higher scores to some students, particularly those from mainland China, than did expert human graders, while systematically underscoring African Americans and, at various points, Arabic, Spanish, and Hindi speakers—even after repeated attempts to reconfigure the system to address these disparities (Heilman et al., 2014; Ramineni &

Williamson, 2018). The bias proves particularly severe for African American students, with E-rater underscoring their essays by an average of 0.81 points on a six-point scale compared to expert human evaluators, while overscoring Chinese students by 1.3 points (Ramineni & Williamson, 2018).

The algorithms used to score student responses are trained on datasets that reflect the preferences and biases of predominantly white, middle-class educators. When students use linguistic patterns, cultural references, or rhetorical styles that differ from these training examples, they receive lower scores regardless of the quality of their ideas or reasoning. African American students consistently receive low marks from automated systems for grammar, style, and organization—metrics closely correlated with essay length—despite often performing substantially better when evaluated by expert human graders (Ramineni & Williamson, 2018).

These systems focus heavily on surface-level features like sentence length, vocabulary, spelling, and subject-verb agreement—precisely the aspects of writing where English language learners and students from linguistically diverse backgrounds are more likely to differ from dominant linguistic norms (Guskin, 2019). The algorithms prove unable to assess more sophisticated aspects of writing such as creativity, critical thinking, or the quality of argumentation, instead prioritizing conformity to standardized linguistic conventions.

Research at the New Jersey Institute of Technology further documented these discriminatory patterns when examining the ACCUPLACER automated scoring system used by the College Board for college placement testing. The study found that the system failed to reliably predict academic success for female, Asian, Hispanic, and African American students, leading the institution to conclude it could not legally defend use of the test if challenged under federal civil rights legislation (Elliot et al., 2012).

These biases compound over time as students receive algorithmic feedback that encourages them to suppress their cultural identities and adopt linguistic patterns associated with dominant groups. The result is an educational system that systematically devalues the intellectual contributions of students from diverse backgrounds while appearing to provide objective assessment of academic ability. As automated essay scoring systems are deployed as primary graders in at least 21 states, with only 5-20% of essays receiving human review, millions of students face assessment by systems that exhibit documented bias against their linguistic and cultural backgrounds (Guskin, 2019).

The Environmental Burden of Educational Surveillance

The technological infrastructure required to support algorithmic surveillance in educational settings creates environmental burdens that disproportionately affect communities of color. Data centers that process surveillance video, store student records, and run behavior analysis algorithms require enormous amounts of energy and cooling infrastructure. These facilities typically rely on electrical grids powered predominantly by fossil fuels, creating environmental costs that are disproportionately concentrated in communities of color where such infrastructure is preferentially located.

Resistance & Alternative Approaches

Some educational institutions have begun recognizing the discriminatory impacts of algorithmic admissions and assessment systems, implementing reforms designed to reduce bias and increase diversity. Test-optional admissions policies, holistic review processes, and alternative assessment methods represent attempts to counter algorithmic discrimination in educational gate-keeping.

However, these reforms often prove insufficient because they fail to address the underlying structural inequalities that create disparities in traditional academic metrics. Students from under-resourced schools may still be disadvantaged by holistic review processes that consider factors like extracurricular activities, volunteer work, and leadership experience that require economic resources and cultural capital.

Research on test-optional admissions policies shows mixed results, with some institutions achieving modest increases in diversity while others see minimal change. The limited impact often reflects underlying structural inequalities that create disparities in academic preparation and extracurricular opportunities that holistic review processes continue to value.

More fundamental reforms require recognizing that educational opportunity cannot be separated from broader patterns of racial and economic inequality. Truly equitable educational systems must address the structural conditions that create disparities in academic preparation while developing assessment and admissions processes that recognize diverse forms of intellectual potential.

Community-based approaches to educational technology governance offer promising alternatives to top-down algorithmic control. When communities have meaningful input into the design and deployment of educational technologies, they can ensure that these systems serve student needs rather than perpetuating patterns of exclusion.

Emerging models of community-controlled educational technology initiatives demonstrate promising alternatives to institutional algorithmic control. These approaches prioritize community input in technology design and deployment, ensuring that systems serve student and family needs rather than perpetuating patterns of exclusion.

Implications for Educational Equity

The proliferation of algorithmic systems in education represents a fundamental shift in how institutions make decisions about student access and opportunity. These systems promise efficiency and objectivity but often serve to mechanize and legitimize discriminatory practices that have historically excluded students from marginalized communities.

Understanding these systems requires recognizing that they operate within broader structures of racial and economic inequality. Algorithmic bias in education is not simply a technical problem that can be solved through better programming or more diverse training data. Instead, it reflects deeper patterns of institutional racism that must be addressed through structural change rather than technological fixes.

The stakes could not be higher. Education remains one of the few pathways for economic mobility in an increasingly stratified society. When algorithmic systems control access to educational opportunities, they shape the life chances of entire generations while appearing to operate according to neutral, scientific principles.

Addressing algorithmic discrimination in education requires both immediate reforms to reduce bias in existing systems and longer-term structural changes that address the inequalities these systems perpetuate. This work cannot be left to technologists alone but requires collaboration between educators, community advocates, policymakers, and students themselves.

Student Assessment & the Algorithmic Measurement of Learning

The deployment of algorithmic assessment systems in K-12 education reveals how educational technology can systematically misinterpret and undervalue the knowledge and capabilities of students from diverse cultural backgrounds. These systems promise more objective and efficient evaluation of student learning, but they often mistake cultural differences for academic deficiencies.

Computer-based assessment systems analyze student responses to test questions using natural language processing and machine learning algorithms that evaluate not just whether answers are correct, but how they are expressed and reasoned. These systems can process thousands of student responses simultaneously, providing immediate feedback and generating detailed analytics about learning progress.

But the algorithmic interpretation of student responses reflects the cultural biases embedded in the training data and assessment frameworks. When algorithms learn to recognize "good" answers from datasets composed primarily of responses from affluent white students, they may systematically undervalue equally valid responses that reflect different cultural knowledge bases or communication styles.

The bias is particularly evident in writing assessment systems that analyze student essays and short-answer responses. These systems evaluate factors like vocabulary usage, sentence structure, and organizational patterns to generate scores that supposedly measure writing quality and critical thinking skills.

However, research by linguists and education scholars has documented how these systems systematically penalize students whose writing reflects African American Vernacular English (AAVE) or other non-standard dialects, even when the content demonstrates sophisticated thinking and communication skills.

The algorithms interpret linguistic diversity as evidence of poor writing ability rather than recognizing different but equally valid approaches to communication.

Similar biases appear in mathematics assessment systems that analyze student problem-solving approaches. These systems are designed to evaluate not just whether students arrive at correct answers, but whether they use appropriate mathematical reasoning and problem-solving strategies.

But the algorithmic evaluation of mathematical reasoning often reflects narrow definitions of appropriate problem-solving approaches that may not recognize culturally specific mathematical knowledge or alternative solution strategies. Students who solve problems using methods learned in community contexts or through different cultural mathematical traditions may be penalized by algorithms that don't recognize the validity of their approaches.

Reading comprehension assessment systems exhibit particularly concerning patterns of bias, especially when evaluating students whose first language is not English or whose cultural backgrounds differ from the populations represented in the training data. These systems analyze student responses to reading passages to assess comprehension, inference-making, and critical analysis skills.

But the algorithmic evaluation often relies on cultural knowledge and interpretive frameworks that may not be accessible to students from diverse backgrounds. Reading passages and questions may assume familiarity with cultural references, social contexts, or interpretive conventions that are specific to dominant cultural groups, leading algorithms to underestimate the reading comprehension abilities of students from different cultural backgrounds.

The cumulative impact of biased assessment systems creates systematic barriers to educational advancement for students from underrepresented communities. When algorithmic assessments consistently underestimate their capabilities, these students are less likely to be identified for gifted and talented programs, advanced coursework, or enrichment opportunities that prepare them for competitive college applications.

The tracking systems that use algorithmic assessment data to make course placement recommendations systematically channel students of color into lower-level academic tracks, reducing their access to the advanced coursework that is increasingly necessary for competitive college admission. Students who demonstrate strong academic potential through other measures may be excluded from advanced programs because algorithmic assessments failed to recognize their capabilities.

Educational Technology Deployment: The Digital Divide in Schools

The deployment of educational technology across American schools follows patterns of digital redlining that systematically advantage affluent, predominantly white schools while limiting access to high-quality educational technology in schools serving predominantly students of color.

This technology deployment disparity operates through multiple mechanisms that compound existing educational inequalities. Schools in affluent districts typically have access to high-speed internet, modern computing devices, and comprehensive technical support that enables effective educational technology implementation. They can afford to purchase premium educational software, provide extensive teacher training, and maintain up-to-date equipment that supports innovative pedagogical approaches.

In contrast, schools in under-resourced districts often lack the basic infrastructure necessary for effective educational technology deployment. Internet connections may be unreliable or insufficient to support multiple simultaneous users, computing devices may be outdated or in poor repair, and technical support may be limited or nonexistent.

These infrastructure disparities create systematic differences in educational opportunities that algorithmic systems help perpetuate and amplify. When schools with better technology infrastructure can provide more sophisticated educational experiences, their students develop stronger digital literacy skills and familiarity with technology systems that advantage them in college applications and career preparation.

The COVID-19 pandemic made these disparities starkly visible when schools shifted to remote learning and students' access to educational technology became crucial for academic continuity. Students in affluent districts typically had access to high-speed internet, modern computing devices, and comprehensive technical support that enabled effective remote learning.

But students in under-resourced districts often lacked reliable internet access, suitable computing devices, or technical support necessary for remote learning participation. The result was systematic exclusion from educational opportunities that further widened achievement gaps and reduced college preparation for students who were already disadvantaged by resource inequities.

The algorithmic systems that managed remote learning platforms and assessed student engagement often misinterpreted the impacts of technology access disparities as evidence of student disengagement or academic deficiency. Students who had difficulty participating in online classes due to technology limitations were flagged by algorithmic monitoring systems as at-risk or disengaged, leading to interventions that focused on individual behavior modification rather than addressing structural barriers to participation.

Educational data mining systems that analyze student engagement with online learning platforms exhibit similar biases, interpreting the impacts of technology access disparities as individual student characteristics rather than recognizing structural inequities. Students who cannot afford high-speed internet or modern computing devices may appear less engaged with online learning materials, leading algorithms to predict lower academic success and recommend less challenging coursework.

The bias extends to the educational software and online learning platforms that schools can afford to purchase. Affluent districts can afford premium educational technology products that offer sophisticated adaptive learning features, comprehensive curriculum coverage, and advanced analytics that support personalized instruction.

Under-resourced districts often rely on free or low-cost educational technology products that may have limited functionality, outdated content, or advertising-supported revenue models that create additional barriers to learning. The algorithmic systems embedded in these lower-quality products may be less sophisticated in their analysis of student learning and more likely to exhibit biases that disadvantage students of color.

Surveillance Technology in Schools: Criminalizing Childhood

The deployment of surveillance technology in American schools has created unprecedented systems of monitoring and control that disproportionately impact students of color while normalizing carceral approaches to education. These systems, marketed as enhancing school safety and improving educational outcomes, actually criminalize normal adolescent behavior and create pathways from school to prison that particularly harm Black and Hispanic students.

School surveillance systems have expanded rapidly since the 1990s, driven by moral panics about school violence and encouraged by federal funding programs that incentivize security technology adoption. Modern school surveillance infrastructure includes security cameras, metal detectors, facial recognition systems, social media monitoring software, and behavioral prediction algorithms that claim to identify students at risk of violence or academic failure.

But research on school surveillance outcomes reveals systematic disparities in how these technologies impact students from different racial backgrounds. Surveillance systems are more likely to be deployed in schools serving predominantly students of color, creating differential exposure to monitoring and control that reflects broader patterns of criminalization in American society.

The algorithmic analysis of surveillance data exhibits systematic biases that interpret normal adolescent behavior differently depending on the race of the student being monitored. Behavioral prediction algorithms trained on school discipline data learn to associate certain behaviors with disciplinary referrals, but the training data reflects decades of discriminatory discipline practices that disproportionately punish students of color for the same behaviors that result in lesser consequences for white students.

When algorithms learn from this biased disciplinary data, they reproduce discriminatory patterns while claiming mathematical objectivity. Students of color are more likely to be flagged by algorithmic systems as exhibiting concerning behavior, leading to interventions that channel them toward disciplinary rather than supportive responses.

The social media monitoring systems deployed in many school districts exhibit particularly concerning patterns of bias. These systems use natural language processing algorithms to analyze student social media posts for content that might indicate risk of violence, self-harm, or other concerning behaviors.

But the algorithmic interpretation of social media content often reflects cultural biases that misinterpret the communication styles and cultural references common in communities of color. Students who use African American Vernacular English or reference cultural contexts unfamiliar to the algorithmic systems may be flagged as exhibiting concerning behavior, even when their posts reflect normal social communication.

The facial recognition systems deployed in some school districts create additional opportunities for discriminatory surveillance. These systems claim to enhance security by identifying individuals who pose threats to school safety, but research on facial recognition accuracy reveals systematic biases that disproportionately misidentify people of color.

When facial recognition systems generate false positive identifications of students of color, these students may face disciplinary consequences or security interventions based on algorithmic errors. The technology creates opportunities for harassment and criminalization that particularly impact students who are already subject to disparate treatment in school discipline systems.

The behavioral prediction algorithms used in some districts to identify students at risk of academic failure or dropout exhibit similar biases. These systems analyze student data including attendance patterns, disciplinary records, and academic performance to generate risk scores that supposedly identify students who need additional support.

But the algorithmic analysis often confuses the impacts of structural inequities with individual student characteristics. Students who miss school due to transportation challenges, family responsibilities, or economic pressures may be flagged as disengaged, leading to interventions that focus on individual behavior change rather than addressing structural barriers to school attendance.

The Promise and Peril of Personalized Learning

Educational technology companies have promoted "personalized learning" as a solution to educational inequality, promising algorithmic systems that can adapt instruction to individual student needs and learning styles. These systems claim to provide customized educational experiences that help all students reach their full potential, regardless of background or previous academic experience.

But the implementation of personalized learning systems often reproduces and amplifies existing educational inequalities while creating new forms of algorithmic discrimination. The artificial intelligence systems that power personalized learning platforms learn from educational data that reflects decades of biased teaching practices and discriminatory assessment systems.

When algorithms learn to recognize "effective" instruction from training data that includes the outcomes of biased educational practices, they may systematically recommend different instructional approaches for students from different racial backgrounds. Students of color may be directed toward more basic, skills-focused instruction while white students receive more challenging, creativity-focused learning experiences.

The algorithmic assessment of student learning preferences and capabilities often reflects cultural biases that mistake differences for deficiencies. Students whose learning approaches don't conform to dominant cultural patterns may be assessed as having learning difficulties or limited academic potential, leading personalized learning systems to provide less challenging instruction.

Research on personalized learning implementations has documented concerning patterns where these systems systematically provide lower-quality educational experiences to students from underrepresented communities. Rather than addressing educational inequality, algorithmic personalization often reinforces existing disparities while claiming to serve individual student needs.

The data collection practices that enable personalized learning systems also raise significant privacy concerns that disproportionately impact students from marginalized communities. These systems collect detailed information about student learning behaviors, social interactions, and academic performance that can be used to make predictions about future educational and career outcomes.

When this detailed student data is shared with colleges, employers, or other institutions, it can perpetuate discrimination by providing algorithmic predictions about student potential that reflect biased assessment systems rather than actual capabilities. Students from underrepresented communities may face additional barriers to educational and career advancement based on algorithmic predictions generated during their K-12 education.

Higher Education's Algorithmic Infrastructure

Universities have become laboratories for algorithmic discrimination, deploying sophisticated data mining and predictive analytics systems that claim to improve student success while actually perpetuating patterns of exclusion. These systems analyze vast amounts of student data to make predictions about academic outcomes and recommendations about interventions, but their operations often reflect the same biases that characterize other educational algorithms.

Student success prediction systems analyze factors like high school GPA, standardized test scores, first-semester grades, and demographic characteristics to identify students at risk of academic failure or dropout. Universities use these predictions to allocate tutoring resources, academic support services, and financial aid in ways that supposedly maximize student success.

But the algorithmic analysis often confuses correlation with causation, interpreting the impacts of structural inequities as individual student characteristics. Students from low-income backgrounds may be predicted to have higher dropout risk not because of individual academic deficiencies, but because they face financial pressures, work responsibilities, and other structural barriers that the algorithms interpret as personal limitations.

The resulting interventions often focus on individual behavior modification rather than addressing the structural barriers that create academic challenges for students from underrepresented communities. Students flagged as high-risk may be directed toward remedial coursework, intensive advising, or other interventions that actually limit their academic advancement rather than providing the support they need to succeed.

University learning management systems collect detailed data about student engagement with online course materials, participation in discussion forums, and completion of assignments. This data is analyzed by algorithmic systems that claim to provide insights into student learning and engagement patterns.

But the algorithmic interpretation of student engagement data often reflects cultural biases and assumptions about appropriate learning behaviors. Students who engage with course materials in ways that don't conform to dominant cultural patterns may be assessed as disengaged or unmotivated, leading to interventions that misunderstand their actual learning needs.

The grade prediction algorithms deployed by some universities exhibit systematic biases that disadvantage students from underrepresented communities. These systems analyze historical grading data to predict student performance in specific courses, but the training data reflects decades of biased grading practices that may systematically underestimate the capabilities of students of color.

When universities use these algorithmic predictions to make decisions about course placement, academic support allocation, or financial aid distribution, they perpetuate discriminatory patterns while claiming to use objective, data-driven approaches.

The Path Forward: Toward Equitable Educational Technology

Addressing algorithmic discrimination in educational technology requires comprehensive approaches that recognize the social and cultural dimensions of learning while ensuring that technological systems serve educational equity rather than perpetuating exclusion.

The priority is algorithmic transparency and accountability in educational systems. Students, families, and educators have a right to understand how algorithmic systems make decisions about academic opportunities, assessment outcomes, and resource allocation. Educational institutions should be required to disclose the factors that algorithmic systems consider, the data sources they use, and the outcomes they produce for different demographic groups.

Algorithmic auditing should be mandatory for educational technology systems, with regular assessments of discriminatory impact and requirements for corrective action when bias is identified. These audits should include community participation, ensuring that the voices of students and families from affected communities are centered in evaluating algorithmic systems.

Educational data privacy protections must be strengthened to prevent the use of student data in ways that perpetuate discrimination. Students should have the right to understand what data is collected about them, how it is used, and who has access to it. They should also have the right to challenge algorithmic decisions that significantly impact their educational opportunities.

Alternative assessment approaches that recognize cultural diversity and multiple forms of knowledge should be developed and implemented. Rather than relying solely on standardized assessments that exhibit systematic bias, educational systems should adopt portfolio-based evaluation, culturally responsive assessment methods, and holistic review processes that recognize the full range of student capabilities.

Community involvement in educational technology governance is essential for ensuring that these systems serve student needs rather than perpetuating exclusion. Students, families, and community members should have meaningful input into decisions about educational technology adoption, implementation, and evaluation.

The development of educational algorithms should include diverse perspectives from the design stage, ensuring that systems are built to recognize and value different forms of knowledge and learning approaches. This requires diversifying the technology workforce and creating meaningful partnerships between technology developers and communities that have been historically excluded from educational opportunities.

The future of educational technology depends on our collective commitment to ensuring that these systems serve all students rather than perpetuating historical patterns of exclusion. This requires not just technical fixes to biased algorithms, but fundamental changes in how we understand the purposes of education and the role of technology in supporting human development.

As algorithmic systems become more sophisticated and pervasive in educational contexts, the stakes of getting this right continue to grow. Whether educational technology becomes a tool for equity or a mechanism for perpetuating discrimination depends on our willingness to center justice and community empowerment in how these systems are designed, implemented, and governed.

The classroom should be a place where all students can develop their full potential, but algorithmic discrimination threatens to turn educational institutions into sorting mechanisms that reproduce existing inequalities with mathematical precision. The choice is ours: we can continue to accept educational technology systems that perpetuate discrimination, or we can demand alternatives that serve the promise of education as a pathway to opportunity and human flourishing.

PART V: RESISTANCE AND REFORM

CHAPTER 10:
COMMUNITY-CENTERED
TECHNOLOGY GOVERNAN

Theoretical Frameworks for Democratic Technology and Environmental Justice

When xAI announced plans to import a massive natural gas power plant to fuel its Memphis data center operations in 2024, the company strategically targeted South Memphis—a historically Black community already bearing 4x the national cancer rate and decades of industrial pollution. Yet within weeks of the announcement, something unprecedented happened: community organizers, environmental justice advocates, and technology researchers coalesced around a shared framework that directly challenged not just this specific project, but the fundamental assumptions underlying corporate technology deployment in marginalized communities.

This resistance represented more than localized opposition to environmental harm—it embodied emerging practices of community-centered technology governance that prioritize democratic participation over corporate extraction, environmental justice over profit maximization, and community sovereignty over technological colonialism. As residents packed city council meetings demanding answers about air quality impacts and energy grid strain, they were simultaneously articulating alternative visions for how artificial intelligence infrastructure could serve rather than exploit their communities.

The Memphis organizing campaign illustrates core principles that distinguish community-centered technology governance from conventional approaches: the centering of affected communities as primary decision-makers, the integration of environmental and social justice analysis into technology policy, and the development of participatory mechanisms that enable meaningful democratic control over technological systems. These principles challenge the prevailing assumption that technological development inevitably requires sacrificing vulnerable communities to corporate interests.

Theoretical Foundations of Democratic Technology Governance

Community-centered technology governance emerges from theoretical traditions that reject technological determinism—the belief that technology develops according to its own internal logic independent of social forces and political choices. Instead, these frameworks recognize technology as fundamentally political, embodying particular values, power relationships, and distributional consequences that reflect the interests of those who control technological development (Winner, 1980).

The Science, Technology, and Society (STS) tradition provides crucial theoretical grounding for understanding how seemingly neutral technical systems encode social relationships and power structures. Technologies are not merely tools applied to social problems but rather sociotechnical systems that co-evolve with social institutions, economic relationships, and political arrangements. This perspective reveals how AI infrastructure deployment patterns reflect and reproduce existing inequalities while creating new forms of digital stratification.

Critical race theory offers additional theoretical resources for analyzing how technology systems perpetuate racial subordination through ostensibly race-neutral mechanisms. The concept of "colorblind racism" proves particularly relevant for understanding how algorithmic systems and infrastructure placement decisions produce racially disparate outcomes while maintaining plausible deniability about discriminatory intent. Community-centered governance frameworks explicitly center race and racism as organizing principles for understanding technological impacts.

Feminist science and technology studies contribute insights about how gendered power relationships shape technological development and deployment. The concept of "situated knowledge" emphasizes that marginalized communities possess unique epistemological perspectives on technological systems based on their experiences of technological harm and exclusion. This framework validates community knowledge as essential for developing more just and effective technology governance approaches.

Environmental Justice and Technology Infrastructure

Environmental justice theory provides essential foundations for community-centered technology governance by revealing how environmental harms concentrate in marginalized communities through systematic patterns of discrimination rather than natural market forces or technical necessity. The environmental justice movement's core principles—meaningful participation, fair treatment, and community self-determination—offer direct templates for democratic technology governance (Bullard, 1990).

The concept of "environmental racism" proves particularly relevant for analyzing AI infrastructure deployment patterns. Research documenting how data centers, surveillance systems, and other technology infrastructure concentrate in communities of color demonstrates that environmental and technological harms operate through interconnected mechanisms of spatial marginalization and resource extraction.

Procedural environmental justice demands that affected communities have meaningful opportunities to participate in decisions affecting their environments. This principle directly challenges technology deployment processes that treat community consultation as public relations rather than genuine democratic participation. Community-centered governance requires that affected populations have decision-making power, not merely consultation opportunities.

Distributive environmental justice examines how environmental benefits and burdens are allocated across different communities. Applied to technology infrastructure, this framework reveals how AI development concentrates benefits (jobs, tax revenue, technological capacity) in privileged communities while externalizing costs (environmental pollution, health impacts, surveillance) to marginalized populations.

Recognition-based environmental justice emphasizes the importance of acknowledging different ways of knowing and valuing environmental resources. For technology governance, this means recognizing community knowledge systems, cultural values, and alternative approaches to technological development that may conflict with dominant corporate models.

Participatory Democracy and Technology Assessment

Participatory democracy theory offers frameworks for reimagining technology governance as a fundamentally democratic process rather than a technical exercise dominated by experts and corporate interests. The concept of "strong democracy" emphasizes citizen participation in substantive decision-making rather than merely periodic electoral input (Barber, 1984).

Participatory technology assessment represents one approach to democratizing technology governance by involving affected communities in evaluating technological systems before, during, and after deployment. This approach recognizes that communities experiencing technological impacts possess essential knowledge about system performance, unintended consequences, and alternative approaches that technical experts may overlook.

The theory of communicative action provides additional resources for understanding how democratic technology governance might function through processes of inclusive deliberation aimed at reaching reasoned consensus about technological choices (Habermas, 1981). However, community organizing traditions emphasize that meaningful participation requires addressing power imbalances that may prevent marginalized voices from being heard in formal deliberative processes.

Critical theories of participatory democracy highlight how seemingly inclusive processes can reproduce existing power relationships if they fail to address structural inequalities. Community-centered technology governance must therefore combine participatory mechanisms with explicit attention to power-building within marginalized communities and redistribution of decision-making authority.

Challenging Environmental Racism in Technology Infrastructure

Communities challenging environmental racism in technology infrastructure deployment have developed sophisticated analytical frameworks for documenting discriminatory patterns and developing alternative approaches. Spatial analysis techniques enable communities to map relationships between technology infrastructure placement and demographic characteristics, revealing patterns that might otherwise remain invisible.

The Memphis xAI case exemplifies how communities can use spatial analysis to challenge corporate claims about site selection. When xAI representatives argued that Memphis offered optimal conditions for data center operations, community researchers documented how the company specifically targeted South Memphis—a predominantly Black area with existing environmental burdens—while avoiding wealthier, whiter areas of the city with similar technical characteristics.

Geographic Information Systems (GIS) mapping enables communities to visualize cumulative environmental impacts by overlaying technology infrastructure with existing pollution sources, health outcomes, and demographic data. These visualizations prove powerful tools for demonstrating environmental racism while building support for alternative approaches to infrastructure development.

Community-based participatory research (CBPR) methods enable residents to conduct their own studies of technology infrastructure impacts rather than relying solely on corporate-sponsored assessments or government studies that may not reflect community priorities. CBPR approaches recognize community members as co-researchers capable of identifying research questions, collecting data, and interpreting results.

Legal and Regulatory Strategies

Communities challenging environmental racism in technology infrastructure have developed multi-faceted legal strategies that combine traditional environmental law with civil rights approaches and emerging technology governance frameworks. Title VI of the Civil Rights Act of 1964 prohibits discrimination in federally funded programs, providing one avenue for challenging infrastructure projects that produce racially disparate impacts.

The National Environmental Policy Act (NEPA) requires environmental impact assessments for major federal actions, creating opportunities for community input on technology infrastructure projects involving federal permits, funding, or land use. However, communities have learned that meaningful participation in NEPA processes requires technical expertise, legal representation, and sustained organizing capacity that may exceed available resources.

State-level environmental justice laws offer additional tools for challenging discriminatory infrastructure deployment. Some states have adopted cumulative impact assessment requirements that consider how new projects contribute to existing environmental burdens rather than treating each project in isolation. These frameworks prove particularly relevant for technology infrastructure that may add to communities' cumulative exposure to air pollution, noise, and other environmental stressors.

Emerging "right to healthy environment" legal frameworks provide constitutional or statutory protections that communities can invoke to challenge harmful technology infrastructure. While these legal tools remain underdeveloped in most U.S. jurisdictions, international human rights frameworks increasingly recognize environmental rights that could apply to technology infrastructure deployment.

Direct Action and Civil Disobedience

Community resistance to environmental racism in technology infrastructure increasingly employs direct action tactics that disrupt business-as-usual while building popular support for alternative approaches. The long tradition of environmental justice organizing provides models for combining confrontational tactics with policy advocacy and community education.

Pipeline resistance campaigns offer instructive examples of how communities can challenge extractive infrastructure through sustained direct action combined with legal strategies and broader political organizing. The fight against the Dakota Access Pipeline demonstrated how indigenous-led resistance could successfully challenge corporate infrastructure projects while articulating alternative visions of energy democracy and environmental protection.

Technology infrastructure resistance faces unique challenges due to the perceived legitimacy of digital development and the technical complexity of AI systems. However, communities have successfully adapted traditional environmental justice tactics to challenge data centers, surveillance systems, and other technology infrastructure through strategies including construction blockades, permit hearing disruptions, and investor pressure campaigns.

Civil disobedience tactics prove particularly effective when combined with broader political education about the connections between technology infrastructure and environmental racism. Community organizers emphasize that resistance to specific projects must connect to larger struggles for democratic control over technological development and environmental protection.

Participatory Design Principles for Community Empowerment

Participatory design frameworks offer alternatives to top-down technology development by centering affected communities as co-designers rather than passive users or impacted populations. These approaches recognize that communities possess essential knowledge about their needs, resources, and priorities that must inform technological system design from the earliest stages.

The principle of "nothing about us, without us" from disability rights organizing applies directly to technology governance: decisions about technological systems affecting communities must include meaningful participation by those communities. This requires moving beyond consultation or user feedback to genuine power-sharing in design decisions, resource allocation, and system governance.

Co-design methodologies enable communities to participate directly in defining problems, developing solutions, and evaluating outcomes rather than merely providing input on predetermined technical approaches. These methods recognize community members as experts on their own experiences and environments rather than treating technical specialists as the sole authorities on technological solutions.

Participatory design principles include: accessibility and inclusion in design processes, transparency about technical choices and their implications, community ownership or control over technological systems, accountability mechanisms that enable ongoing community oversight, and sustainability approaches that consider long-term community impacts rather than short-term corporate profits.

Community Technology Networks

Community technology networks represent practical applications of participatory design principles through locally controlled infrastructure that serves community needs rather than corporate interests. These networks demonstrate how communities can develop technological capacity independently while maintaining democratic control over technological development.

Mesh networking technologies enable communities to create communication infrastructure that operates independently of corporate internet service providers while remaining under community control. Examples include Red Hook WiFi in Brooklyn, which provided essential communication services during Hurricane Sandy, and indigenous communities in Oaxaca, Mexico, that have developed community-controlled cellular networks.

Community broadband networks offer alternatives to corporate internet service by treating connectivity as a public utility rather than a private commodity. Municipal broadband initiatives demonstrate how communities can provide high-quality internet access while maintaining democratic accountability and community control over technological infrastructure.

Community-owned renewable energy projects provide models for democratic ownership of technology infrastructure that serves community needs while generating economic benefits for local residents. These projects demonstrate how communities can develop technological capacity while maintaining environmental sustainability and democratic governance.

Technology Assessment and Community Oversight

Participatory technology assessment enables communities to evaluate technological systems according to their own values and priorities rather than accepting corporate or government assessments of technological impacts. These approaches recognize that different communities may reach different conclusions about technological benefits and risks based on their particular circumstances and values.

Community-controlled technology assessment processes typically include: community education about technological systems and their implications, inclusive deliberation about community values and priorities, independent technical analysis that reflects community concerns, evaluation of alternatives to proposed technological approaches, and ongoing monitoring of technological impacts after deployment.

Consensus-building processes enable communities to develop shared positions on technological issues while respecting diverse perspectives within communities. These processes require significant time and resources but produce more durable agreements than top-down decision-making approaches that may generate ongoing conflict.

Community oversight mechanisms enable ongoing democratic control over technological systems after initial deployment decisions. These may include community advisory boards with real decision-making power, regular public reporting requirements, community-controlled evaluation processes, and mechanisms for modifying or removing technological systems that fail to serve community needs.

Indigenous Data Sovereignty and Alternative Governance Models

Indigenous data sovereignty emerges from broader frameworks of indigenous self-determination and sovereignty that challenge colonial control over indigenous peoples, territories, and knowledge systems. The concept asserts that indigenous peoples have inherent rights to control data about their communities, including collection, ownership, and application of data in ways that serve indigenous community needs and values (Tuck & Yang, 2012).

Traditional indigenous governance systems provide alternative models for technology governance that prioritize collective decision-making, intergenerational responsibility, and ecological sustainability rather than individual ownership, short-term profits, and resource extraction. These systems demonstrate that democratic technology governance can operate according to fundamentally different principles than dominant Western approaches.

The principle of "indigenous data governance" extends beyond data sovereignty to encompass broader questions about how technological systems affect indigenous communities and how indigenous peoples can maintain control over technological development in their territories. This includes questions about AI systems that impact indigenous lands, surveillance technologies that monitor indigenous activities, and digital platforms that represent indigenous cultures.

Legal frameworks for indigenous data sovereignty are emerging through tribal law, international human rights instruments, and collaborative agreements between indigenous communities and research institutions or technology companies. The United Nations Declaration on the Rights of Indigenous Peoples provides international legal grounding for indigenous control over information and technology affecting indigenous communities.

Community Ownership Models

Indigenous communities have developed innovative ownership models for technological infrastructure that maintain community control while enabling technological development. These models challenge dominant assumptions about private property and corporate ownership by treating technology as a community resource rather than private commodity.

Tribal ownership of technology infrastructure enables indigenous communities to develop technological capacity while maintaining sovereignty over technological systems. Examples include tribal internet service providers that serve reservation communities, community-owned renewable energy projects that reduce dependence on external utilities, and tribally controlled data centers that provide technology services under indigenous governance.

Community land trusts provide models for collective ownership that could apply to technology infrastructure. These legal structures enable communities to maintain permanent ownership of land and infrastructure while allowing for individual use rights and democratic governance of community resources.

Cooperative ownership models enable communities to share ownership and control of technology infrastructure while maintaining democratic decision-making processes. Technology cooperatives demonstrate how communities can pool resources to develop technological capacity while avoiding dependency on external corporate or government providers.

Commons-based approaches treat technological infrastructure as a shared resource that belongs to the community as a whole rather than private owners. Digital commons initiatives demonstrate how communities can develop technological resources collaboratively while maintaining open access and democratic governance.

Traditional Ecological Knowledge and Technology

Traditional Ecological Knowledge (TEK) systems offer alternative approaches to technology development that prioritize ecological sustainability, community health, and intergenerational responsibility. These knowledge systems demonstrate how technological development can serve community needs while maintaining ecological balance and cultural integrity.

Indigenous science traditions provide different epistemological foundations for understanding technology and its relationship to human and natural communities. These traditions often emphasize reciprocal relationships between humans and technology rather than domination and control, holistic understanding of technological systems rather than reductionist approaches, and long-term thinking about technological impacts rather than short-term optimization.

Biomimicry and ecological design principles emerging from indigenous knowledge traditions offer alternatives to energy-intensive AI systems that prioritize efficiency and sustainability. These approaches demonstrate how technological development can learn from natural systems rather than attempting to dominate or replace them.

Community-based monitoring systems that integrate traditional ecological knowledge with contemporary technology demonstrate how communities can develop technological capacity while maintaining cultural traditions and ecological sustainability. These systems often prove more effective than top-down monitoring approaches because they integrate local knowledge with technical capacity.

Building Grassroots Power to Resist Digital Colonialism

Building grassroots power to resist digital colonialism requires popular education approaches that help community members understand the connections between technology systems and broader patterns of exploitation and oppression. Popular education differs from traditional educational approaches by starting with community experiences and developing critical analysis through collective dialogue and action.

Technology literacy programs that combine technical skills with critical analysis enable community members to understand how technological systems operate while developing political consciousness about technological impacts. These programs recognize that technical knowledge and political analysis must develop together to enable effective resistance to technological oppression.

Community research projects enable residents to investigate technological systems affecting their communities while building analytical capacity and collective knowledge. Examples include community-led studies of surveillance system deployment, air quality monitoring around data centers, and documentation of technology industry impacts on housing costs and community displacement.

Storytelling and cultural work play essential roles in building consciousness about digital colonialism by helping community members connect their individual experiences to broader patterns of technological oppression. Digital storytelling projects enable communities to document their experiences with technology systems while building shared understanding of resistance strategies.

Coalition Building and Movement Connections

Effective resistance to digital colonialism requires building coalitions that connect technology justice organizing with broader social justice movements. These coalitions enable communities to leverage shared resources, coordinate strategies, and develop political analysis that addresses multiple forms of oppression simultaneously.

Environmental justice coalitions provide natural partnerships for communities challenging technology infrastructure because of shared concerns about environmental racism, community health, and democratic participation in environmental decision-making. These coalitions enable technology justice organizers to learn from decades of environmental organizing experience while contributing analysis about technological impacts.

Labor organizing offers additional coalition opportunities because technology industry workers and affected communities often share interests in democratic control over technological development, workplace safety and health protections, and economic justice. Worker-community alliances demonstrate how different constituencies can unite around shared demands for technology justice.

Housing justice coalitions prove particularly relevant in areas experiencing technology industry-driven gentrification and displacement. These coalitions enable communities to address how technology development affects housing costs and community stability while developing shared strategies for community-controlled development.

Anti-surveillance coalitions bring together communities experiencing police surveillance, immigration enforcement, and corporate data collection to develop shared analysis and resistance strategies. These coalitions demonstrate how different forms of surveillance interconnect and how communities can resist technological monitoring through collective action.

Electoral and Policy Strategies

Building grassroots power requires developing electoral and policy strategies that enable communities to influence technology governance through formal political processes while maintaining independent organizing capacity. These strategies must balance engagement with existing political institutions with efforts to transform those institutions to better serve community needs.

Municipal broadband campaigns demonstrate how communities can use local electoral processes to challenge corporate control over technology infrastructure. These campaigns often combine electoral work with community education and direct action to build support for community-controlled alternatives to corporate internet service providers.

Community benefits agreements provide mechanisms for communities to negotiate directly with technology companies about development projects while building community organizing capacity. These agreements demonstrate how communities can leverage corporate development projects to secure community benefits while maintaining ongoing organizing efforts.

Technology policy advocacy enables communities to influence federal and state policy affecting technology systems while building political capacity for broader technology justice organizing. Examples include campaigns for stronger privacy protections, algorithmic accountability requirements, and community oversight of surveillance technologies.

Electoral strategies must connect to broader movement building efforts rather than substituting electoral work for community organizing. Successful technology justice campaigns typically combine electoral advocacy with direct action, community education, and grassroots power building to create multiple pressure points for social change.

Economic Alternatives and Community Development

Resistance to digital colonialism requires developing economic alternatives that enable communities to meet their technological needs without depending on extractive corporate relationships. These alternatives demonstrate how communities can develop technological capacity while maintaining democratic control and community ownership.

Community Development Financial Institutions (CDFIs) provide models for community-controlled financing of technology infrastructure that serves community needs rather than corporate profits. These institutions enable communities to pool resources for technology development while maintaining democratic governance and community accountability.

Social economy approaches that combine market mechanisms with social values offer alternatives to purely profit-driven technology development. Examples include technology cooperatives, community-owned enterprises, and social ventures that use business methods to achieve community development goals.

Time banking and mutual aid networks enable communities to share technology resources and skills without relying on corporate markets or government services. These systems demonstrate how communities can meet their technological needs through cooperation and reciprocity rather than commercial exchange.

Community ownership of technology infrastructure enables communities to capture economic benefits from technology development while maintaining control over technological systems. Examples include community-owned data centers, renewable energy projects, and broadband networks that generate revenue for community development while serving community needs.

Institutional Transformation and Scaling Resistance

Policy Innovation and Institutional Change

Community-centered technology governance requires transforming existing institutions while creating new institutional forms that better serve community needs. This transformation process must balance working within existing systems with developing alternative approaches that challenge dominant institutional arrangements.

Participatory budgeting processes enable communities to allocate public resources for technology infrastructure according to community priorities rather than technocratic decision-making processes. These processes demonstrate how democratic participation can improve technology planning while building community capacity for ongoing oversight and governance.

Community oversight boards with real decision-making power provide mechanisms for ongoing community control over technology systems rather than one-time consultation processes. These boards must have adequate resources, technical support, and legal authority to function effectively as community governance institutions.

Public ownership models for technology infrastructure offer alternatives to both corporate ownership and traditional government provision by treating technology as a public good under democratic control. Municipal broadband, public banking, and community land trusts provide models for public ownership that maintains democratic accountability.

Regional cooperation approaches enable communities to coordinate technology governance across jurisdictional boundaries while maintaining local democratic control. Examples include regional broadband authorities, multi-jurisdictional environmental justice coalitions, and interstate compacts for technology regulation.

Replicating and Scaling Community Models

Successful community-centered technology governance initiatives face challenges in replicating and scaling their approaches while maintaining democratic participation and community control. Scaling strategies must balance the benefits of coordination and resource sharing with the need to maintain local autonomy and responsiveness to specific community needs.

Network models enable communities to share resources, strategies, and knowledge while maintaining local control over technology governance decisions. Examples include networks of community broadband providers, coalitions of environmental justice organizations, and associations of community land trusts that provide mutual support without centralizing decision-making.

Technical assistance and capacity building programs enable experienced communities to support others developing community-centered technology governance approaches. These programs must respect local autonomy while providing useful resources and avoiding the imposition of outside models on local communities.

Policy advocacy for supporting infrastructure enables communities to secure resources for community-centered technology governance while maintaining independence from government or corporate control. Examples include funding programs for community broadband, technical assistance for community organizations, and regulatory frameworks that support community ownership.

Movement building approaches that connect local community governance initiatives to broader social movements enable communities to coordinate resistance to digital colonialism while maintaining local focus and accountability. These approaches recognize that sustainable change requires transformation at multiple scales simultaneously.

Toward Democratic Technology Futures

Community-centered technology governance represents more than resistance to harmful technology deployment—it articulates alternative visions of how technological development could serve human flourishing and ecological sustainability rather than corporate profit and social control. The frameworks, strategies, and practices examined in this chapter demonstrate that democratic alternatives to corporate technology governance are both necessary and possible.

The Memphis organizing campaign against xAI's power plant import illustrates how communities can successfully challenge specific technology projects while building broader movements for technology justice. When residents packed city council meetings demanding environmental impact assessments and community oversight, they were simultaneously practicing democratic technology governance and building capacity for ongoing resistance to digital colonialism.

However, individual victories against harmful technology projects cannot substitute for broader transformation of the systems that produce technological injustice. Building democratic alternatives requires sustained engagement with the theoretical frameworks, organizational strategies, and institutional innovations that enable communities to assert control over technological development rather than merely reacting to corporate initiatives.

The theoretical frameworks examined in this chapter—environmental justice, participatory democracy, indigenous sovereignty, and critical technology studies—provide essential foundations for understanding how technological systems reproduce and challenge existing power relationships. These frameworks enable communities to develop strategic approaches that address both immediate harms and underlying structural causes of technological injustice.

The practical strategies documented in this chapter—community research, direct action, participatory design, cooperative ownership, and electoral engagement—demonstrate how communities can build power to influence technology governance while developing alternative approaches to technological development. These strategies must be adapted to local conditions while maintaining connections to broader movements for social and economic justice.

Perhaps most importantly, the examples of successful community-centered technology governance examined in this chapter reveal that democratic alternatives to corporate control are not utopian fantasies but practical possibilities that communities are implementing across diverse contexts. From indigenous data sovereignty initiatives to community broadband networks to environmental justice campaigns against harmful infrastructure, communities are demonstrating that technology can serve community needs when communities control technological development.

The urgency of climate change, the acceleration of artificial intelligence development, and the intensification of economic inequality make the development of democratic technology governance approaches essential for human survival and flourishing. The choice between corporate-controlled technology that serves profit over people and community-controlled technology that serves human needs and ecological sustainability remains ours to make—but only if communities organize to claim democratic control over technological development before corporate concentration makes such alternatives impossible.

The next chapter will examine how international cooperation and policy innovation can support community-centered technology governance while addressing the global dimensions of technology justice. The local organizing and alternative development approaches examined here provide essential foundations for

broader transformation, but scaling democratic technology governance requires addressing national and international systems
that currently privilege corporate interests over community
needs.

CHAPTER 11: REGULATORY RESPONSES AND THEIR LIMITAT

International Frameworks for Technology Justice and Environmental Protection

When Congress held its first major hearing on artificial intelligence regulation in May 2023, featuring OpenAI CEO Sam Altman as the star witness, the spectacle revealed both the promise and profound limitations of regulatory responses to AI development. Senators alternated between expressing genuine concern about AI's societal impacts and deferring to industry expertise about technical solutions, while fundamental questions about community control, environmental justice, and democratic governance remained largely unaddressed. The hearing epitomized a regulatory approach that treats AI as a technical problem requiring expert management rather than a political challenge demanding democratic oversight and community empowerment.

This pattern of regulatory capture and democratic deficit characterizes not only American responses to AI development but international efforts that consistently prioritize industry concerns over community needs. While regulatory frameworks like the European Union's AI Act and General Data Protection Regulation (GDPR) represent meaningful advances in technology governance, they remain fundamentally inadequate for addressing the systemic inequalities and environmental harms documented throughout this analysis. More promising approaches emerge from Global South resistance movements and community-led initiatives that challenge the basic assumptions underlying dominant regulatory paradigms.

The limitations of current regulatory responses reflect deeper contradictions within liberal democratic systems that promise inclusive governance while structurally privileging corporate interests over community needs. Examining these limitations alongside emerging alternatives reveals both the necessity and possibility of more democratic approaches to technology governance that center environmental justice, community sovereignty, and global economic equality.

Congressional AI Regulation: Theater and Technocracy

The Spectacle of Expert Consultation

Congressional hearings on AI regulation during 2023-2024 demonstrated how democratic institutions can simulate public engagement while systematically excluding affected communities from meaningful participation in technology governance decisions. The Senate Judiciary Committee's hearing, "Oversight of A.I.: Rules for Artificial Intelligence," featured testimony from Sam Altman, IBM's Christina Montgomery, and NYU professor

Gary Marcus—a panel that perfectly embodied the expert-dominated approach characterizing American AI regulation efforts.

Altman's testimony revealed the sophisticated strategies tech companies employ to shape regulatory discussions while maintaining developmental autonomy. His calls for federal licensing of AI systems and international coordination on AI safety appeared to embrace regulation while actually proposing industry-controlled oversight mechanisms that would legitimize existing power structures. When Senator Richard Blumenthal opened the hearing with an AI-generated speech mimicking his own voice, the demonstration served more to mystify AI capabilities than illuminate governance challenges facing affected communities.

The hearing format itself—expert testimony followed by brief questioning periods—structurally privileged technical expertise over community knowledge while treating complex political questions as narrow technical problems amenable to expert solutions. Missing from these discussions were voices from communities experiencing AI-driven surveillance, environmental harm from data centers, or economic displacement from algorithmic automation. This exclusion reflects broader patterns within American technology policy that treats democratic participation as an obstacle to efficient governance rather than a foundational requirement for legitimate authority.

Subsequent hearings followed similar patterns, featuring industry executives, academic researchers, and policy experts while systematically excluding community organizers, environmental justice advocates, and representatives from affected populations. The House Science Committee's hearing on "Artificial Intelligence: Advancing Innovation Towards the National Interest" exemplified this approach by focusing exclusively on maintaining American competitive advantage rather than addressing domestic impacts on vulnerable communities.

Legislative Initiatives and Corporate Influence

Congressional legislative responses to AI development reveal how corporate influence shapes regulatory frameworks to serve industry interests while creating the appearance of meaningful oversight. The bipartisan AI Task Force, co-chaired by Representatives Jay Obernolte and Ted Lieu, produced recommendations that consistently prioritized industry concerns about innovation and competitiveness over community demands for democratic oversight and environmental protection.

The proposed "CREATE AI Act" exemplifies this regulatory approach by establishing federal coordination mechanisms for AI research and development while avoiding binding requirements for community consultation, environmental impact assessment, or democratic oversight. The legislation treats AI development as an inherently beneficial activity requiring government support rather than a potentially harmful process demanding community control and environmental safeguards.

Industry lobbying expenditures on AI-related issues increased dramatically during 2023-2024, with major technology companies spending millions to influence regulatory discussions. OpenAI's lobbying spending increased from zero in 2022 to over $760,000 in 2023, while established companies like Google, Microsoft, and Meta maintained substantial influence operations focused on shaping AI regulation. This lobbying infrastructure enables companies to participate directly in regulatory development while communities lack comparable resources for sustained engagement.

The revolving door between technology companies and regulatory agencies further compromises the independence of regulatory responses. Former government officials regularly transition to high-paying industry positions, while companies recruit former regulators to navigate government relations. This personnel interchange creates structural conflicts of interest that bias regu-

latory development toward industry preferences rather than public interest considerations.

Algorithmic Accountability Legislation

The Algorithmic Accountability Act, introduced multiple times
in Congress but never enacted, illustrates both the potential and
limitations of legislative approaches to AI governance. The bill
would require companies to assess algorithmic systems for bias,
discrimination, and privacy violations—representing meaningful
progress toward algorithmic transparency and corporate accountability.

However, the legislation's impact assessment requirements focus primarily on individual discrimination rather than structural
inequalities or community-level harms. The bill's emphasis on
bias detection and mitigation treats algorithmic fairness as a
technical problem solvable through better data and improved
methods rather than a political challenge requiring redistribution
of power and resources.

The legislation's enforcement mechanisms rely primarily on
Federal Trade Commission oversight rather than community empowerment or democratic governance structures. This approach
treats affected communities as beneficiaries of regulatory protection rather than agents capable of governing technological
systems affecting their lives. While FTC enforcement could provide valuable accountability mechanisms, it cannot substitute for
community control over technology deployment and governance.

Moreover, the bill's focus on large-scale algorithmic systems
may miss infrastructure deployment decisions that significantly
impact community health and environmental quality. Data center siting, surveillance system installation, and other technology
infrastructure choices operate through different decision-making
processes that may not trigger algorithmic assessment requirements despite their substantial community impacts.

International Regulatory Frameworks: Advances and Limitations

The European Union's Artificial Intelligence Act, enacted in 2024, represents the world's most comprehensive attempt to regulate AI development through binding legal requirements rather than voluntary industry guidelines. The legislation establishes risk-based categories for AI systems, with "unacceptable risk" applications banned entirely and "high-risk" systems subject to extensive compliance requirements including conformity assessments, risk management systems, and human oversight mechanisms.

The Act's prohibition on AI systems for social scoring and real-time facial recognition in public spaces demonstrates how regulatory frameworks can establish meaningful limits on harmful AI applications. These restrictions reflect European values emphasizing privacy rights and limiting state surveillance power—values that contrast sharply with American approaches that often defer to security arguments for expanding surveillance capabilities.

However, the legislation's risk-based approach maintains significant gaps that limit its effectiveness for addressing systemic inequalities and environmental harms. The Act focuses primarily on protecting individual rights rather than addressing collective impacts on communities or ecosystems. Infrastructure deployment decisions, energy consumption patterns, and cumulative environmental effects receive minimal attention despite their substantial implications for environmental justice and community health.

The Act's enforcement mechanisms rely heavily on national regulatory authorities rather than affected communities, limiting democratic participation in ongoing AI governance. While the legislation includes requirements for stakeholder consultation, these provisions treat community input as advisory rather than decision-making participation. This expert-dominated approach may improve AI system performance while maintaining exclusionary governance structures that privilege technical expertise over community knowledge.

The legislation's focus on AI systems rather than broader technology infrastructure means that data center placement, network deployment, and other infrastructure decisions that significantly impact communities may fall outside regulatory scope. This narrow focus reflects the Act's origins in computer science and technology policy communities rather than environmental justice or community organizing traditions.

GDPR as a Model for Technology Governance

The General Data Protection Regulation, implemented in 2018, established important precedents for technology regulation that prioritizes individual rights over corporate convenience while demonstrating how regulatory frameworks can influence global technology development practices. GDPR's requirements for explicit consent, data portability, and deletion rights created new obligations for technology companies while providing individuals with greater control over personal information.

The regulation's extraterritorial reach demonstrates how comprehensive regulatory frameworks can influence global technology practices even when adopted by single jurisdictions. Companies operating globally often implement GDPR-compliant practices worldwide rather than maintaining separate systems for different regulatory environments, suggesting that strong regulatory frameworks can produce broader impacts than their formal jurisdiction might suggest.

However, GDPR's individualistic approach limits its effectiveness for addressing collective harms and systemic inequalities produced by technology systems. The regulation's focus on personal data protection provides minimal tools for addressing algorithmic discrimination, environmental impacts, or community-level surveillance that may not involve personal data collection but nonetheless produce significant social harms.

The regulation's enforcement through national data protection authorities rather than affected communities limits democratic participation while creating inconsistent implementation across member states. While some data protection authorities have pursued aggressive enforcement against major technology companies, others have proven more deferential to industry interests, reducing the regulation's overall effectiveness.

GDPR's compliance requirements impose significant burdens on smaller organizations while larger companies can more easily absorb compliance costs, potentially concentrating market power among established technology giants. This outcome illustrates how regulatory frameworks designed to constrain corporate power may inadvertently strengthen dominant companies while creating barriers for community-controlled alternatives.

Digital Services Act and Platform Governance

The EU's Digital Services Act (DSA), which came into force in 2024, represents another significant advancement in technology regulation by establishing binding requirements for content moderation, algorithmic transparency, and risk assessment on digital platforms. The legislation requires large platforms to provide users with information about content recommendation systems while prohibiting certain forms of targeted advertising to minors.

The Act's requirement for independent audits of platform risk management systems creates new accountability mechanisms that could improve platform governance practices. These audits must assess how platforms address systemic risks including illegal content distribution, negative effects on civic discourse, and impacts on public health and safety.

However, the DSA maintains the fundamental structure of platform governance as a corporate responsibility rather than a democratic process involving affected communities. While the legislation requires platforms to consult with external stakeholders, these consultation requirements stop short of providing communities with decision-making power over platform governance decisions affecting their members.

The Act's focus on content moderation and platform practices may miss broader questions about technology infrastructure, environmental impacts, and economic effects that significantly influence community wellbeing. Like other EU technology regulations, the DSA reflects its origins in competition policy and consumer protection rather than environmental justice or community empowerment traditions.

Global Case Studies: Infrastructure Regulation and Environmental Justice

Nordic Models: Environmental Assessment and Community Consultation

Nordic countries have developed some of the world's most comprehensive frameworks for environmental assessment of technology infrastructure, combining rigorous technical analysis with meaningful community consultation processes. Sweden's environmental impact assessment requirements for large data centers include detailed analysis of energy consumption, cooling requirements, and cumulative environmental effects, while consultation processes provide affected communities with substantial input opportunities.

Norway's approach to data center regulation demonstrates how environmental justice considerations can be integrated into technology infrastructure governance. The country's requirements for renewable energy sourcing and environmental impact mitigation have influenced data center design while consultation processes enable affected communities to influence siting and operational decisions.

However, these frameworks remain limited by their focus on procedural requirements rather than community empowerment or democratic governance. While Nordic consultation processes provide more meaningful participation opportunities than American or many other international approaches, they typically treat community input as advisory rather than binding on infrastructure decisions.

The effectiveness of Nordic approaches depends heavily on robust civil society organizations and high levels of civic engagement that may not exist in other contexts. Transferring these models to countries with weaker environmental movements or less participatory political cultures may require additional institutional innovations that support community organizing and democratic participation.

German Energy Democracy and Infrastructure Planning

Germany's Energiewende (energy transition) program provides instructive examples of how democratic participation can be integrated into large-scale infrastructure planning, though with mixed results for environmental justice and community empowerment. The program's emphasis on renewable energy development and citizen participation in energy planning demonstrates possibilities for more democratic approaches to technology infrastructure.

Community energy cooperatives play significant roles in German renewable energy development, enabling local ownership and democratic governance of energy infrastructure. These cooperatives demonstrate how communities can develop technological capacity while maintaining democratic control over infrastructure serving community needs.

However, the Energiewende's emphasis on technical efficiency and economic optimization has often overwhelmed community participation processes, leading to infrastructure development that serves national energy goals while imposing costs on local communities. Wind farm development has generated significant local opposition in some areas, illustrating tensions between environmental goals and community self-determination.

The program's complex planning processes require substantial technical expertise and legal resources that may exceed the capacity of many affected communities, potentially excluding marginalized voices from meaningful participation. This dynamic illustrates how seemingly inclusive participation processes can reproduce existing inequalities if they fail to address structural barriers to community engagement.

South Korean Digital New Deal and Infrastructure Investment

South Korea's Digital New Deal, launched in 2020, represents one of the world's largest government investments in digital infrastructure development. The program includes substantial funding for 5G networks, artificial intelligence development, and digital government services while emphasizing environmental sustainability and social inclusion goals.

The program's approach to infrastructure planning includes environmental impact assessment requirements and community consultation processes, though these mechanisms provide limited community control over infrastructure decisions. Siting decisions for major technology infrastructure typically prioritize technical and economic factors over community preferences or environmental justice considerations.

South Korea's experience illustrates both possibilities and limitations of government-led technology infrastructure development. While public investment enables more comprehensive planning and environmental consideration than purely market-driven approaches, maintaining democratic accountability requires ongoing community organizing and political engagement that may be difficult to sustain over long development timelines.

The program's emphasis on international competitiveness and technological leadership reflects broader tensions within technology policy between national economic goals and community environmental justice that appear across different political and economic systems.

Global South Resistance and Alternative Models

India's Data Localization and Digital Sovereignty

India's approach to data governance demonstrates how developing countries can assert sovereignty over technology systems while challenging the dominance of Western technology companies. The country's data localization requirements mandate that certain categories of data be stored within national borders, reducing dependence on foreign technology infrastructure while building domestic technological capacity.

The Personal Data Protection Bill, though not yet enacted, would establish comprehensive privacy protections that prioritize national sovereignty over individual rights as conceptualized in European frameworks. This approach reflects broader efforts by developing countries to assert control over technology governance rather than accepting frameworks developed by wealthier nations.

However, India's technology governance approach prioritizes state control over democratic participation, potentially reproducing authoritarian relationships at the national level while challenging international technology dominance. The country's use of technology for surveillance and social control illustrates how digital sovereignty can serve state power rather than community empowerment.

Community resistance to technology infrastructure projects in India demonstrates how local organizing can challenge both corporate and state power over technology deployment. Protests against data center development in rural areas often combine environmental concerns with broader critiques of development models that prioritize economic growth over community wellbeing and environmental protection.

Brazilian Environmental Protection and Indigenous Rights

Brazil's constitutional framework for indigenous rights and environmental protection provides important tools for challenging harmful technology infrastructure, though enforcement remains inconsistent and politically contested. Constitutional requirements for indigenous consultation and consent regarding development projects in indigenous territories create legal mechanisms for community control over technology infrastructure affecting indigenous lands.

The country's environmental licensing processes require impact assessments for large infrastructure projects, including data centers and telecommunications facilities. These processes provide opportunities for community input and environmental protection, though their effectiveness depends heavily on political support for environmental enforcement.

Indigenous communities in Brazil have successfully challenged telecommunications and energy infrastructure projects using constitutional rights frameworks combined with direct action and international advocacy. These campaigns demonstrate how community organizing can leverage legal tools while building broader political support for environmental protection and indigenous sovereignty.

However, political changes can dramatically alter the effectiveness of environmental and indigenous rights protections, as

demonstrated by policy reversals during the Bolsonaro administration (2019-2023). This experience illustrates the importance of
building community power that can resist policy changes while
maintaining environmental and social protections across different political contexts.

African Continental Approaches to Technology Governance

Several African countries have developed innovative approaches to technology governance that prioritize national development goals while addressing concerns about technological
dependence and environmental impacts. Rwanda's approach to
technology infrastructure emphasizes environmental sustainability and social inclusion while building domestic technological
capacity through education and skills development programs.

The African Union's Continental Data Policy Framework represents an ambitious attempt to coordinate technology governance across the continent while asserting African sovereignty
over data and digital development. The framework emphasizes
African ownership and control of technology infrastructure while
promoting continental integration and cooperation.

However, limited institutional capacity and resource constraints often compromise the implementation of progressive
technology policies in African contexts. Many countries lack the
regulatory expertise, enforcement capacity, and technical infrastructure necessary to effectively govern technology systems
while challenging corporate or foreign government power.

Community resistance movements across Africa demonstrate
alternative approaches to technology governance that prioritize
local needs and environmental protection over corporate profits
or national development goals. These movements often combine
traditional governance systems with contemporary organizing

strategies to challenge harmful technology infrastructure while articulating alternative development visions.

Latin American Digital Rights and Community Networks

Latin American countries have developed some of the world's most progressive frameworks for digital rights and community technology governance, reflecting strong traditions of community organizing and social movement activism. Argentina's digital rights legislation includes strong privacy protections and community participation requirements for technology infrastructure development.

Community networking initiatives across Latin America demonstrate how marginalized communities can develop technology infrastructure independently while maintaining democratic governance and community ownership. Examples include indigenous telecommunications networks in Mexico, community broadband projects in rural Colombia, and cooperative internet service providers in urban Argentina.

These community-controlled initiatives often receive support from national and regional policy frameworks that recognize community networking as a legitimate approach to extending technology access. This support enables community organizations to access funding and technical assistance while maintaining autonomy over technology governance decisions.

However, community networking initiatives face ongoing challenges from telecommunications companies and government agencies that may prefer centralized, corporate-controlled infrastructure. Sustaining community technology projects requires ongoing organizing work and political engagement that may strain community resources while competing with other urgent needs.

Successful Environmental Justice Models in Technology Deployment

Costa Rica's Environmental Leadership and Technology Sustainability

Costa Rica has emerged as a global leader in environmental protection and sustainable development, providing models for how technology infrastructure can be developed within ecological limits while prioritizing community wellbeing over economic growth. The country's constitutional right to a healthy environment creates legal foundations for challenging harmful technology infrastructure while its payment for ecosystem services programs demonstrate how environmental protection can be economically sustainable.

The country's approach to data center development emphasizes renewable energy requirements and environmental impact mitigation while consultation processes provide meaningful community input opportunities. Costa Rica's abundant renewable energy resources enable technology infrastructure that operates with minimal environmental impact while supporting national development goals.

However, Costa Rica's small size and specific geographic conditions may limit the transferability of its environmental protection model to larger countries with different ecological and political contexts. The country's success depends partly on tourism revenues that create economic incentives for environmental protection that may not exist in other contexts.

Community participation in environmental decision-making remains limited despite strong legal frameworks, illustrating ongoing challenges in translating environmental rights into democratic governance practices. Building more participatory approaches to technology governance requires continued organizing work and institutional innovation that goes beyond legal protections.

Denmark's Participatory Technology Assessment

Denmark has pioneered participatory technology assessment approaches that involve citizens directly in evaluating emerging technologies before widespread deployment. The Danish Board of Technology has conducted citizen panels, consensus conferences, and deliberative polling exercises that provide ordinary citizens with opportunities to influence technology policy decisions.

These participatory processes have influenced Danish positions on biotechnology, nanotechnology, and artificial intelligence development by incorporating citizen concerns about social and ethical implications alongside technical and economic considerations. The Danish approach demonstrates how technology assessment can be democratized while maintaining technical rigor and policy relevance.

However, participatory technology assessment processes face limitations in addressing structural inequalities and power imbalances that may prevent marginalized voices from meaningful participation. These processes often involve citizens who are already politically engaged rather than those most affected by technology deployment decisions.

The influence of participatory technology assessment on actual policy decisions varies considerably depending on political contexts and institutional arrangements. While these processes can inform policy discussions, they cannot substitute for ongoing community organizing and democratic governance structures that provide affected communities with real decision-making power.

New Zealand's Treaty of Waitangi and Indigenous Technology Rights

New Zealand's Treaty of Waitangi provides constitutional foundations for indigenous participation in technology governance decisions affecting Māori communities and territories. The treaty's principles of partnership, participation, and protection create obligations for government consultation with Māori communities regarding technology infrastructure and data governance.

Māori data sovereignty initiatives demonstrate how indigenous communities can assert control over data and technology systems affecting their communities while maintaining cultural protocols and governance traditions. These initiatives provide models for combining indigenous governance systems with contemporary technology governance challenges.

The country's Privacy Act includes specific provisions for Māori data governance that recognize collective rights alongside individual privacy protections. This approach demonstrates how technology governance frameworks can accommodate different cultural values and governance traditions rather than imposing uniform approaches across diverse communities.

However, the effectiveness of treaty-based protections depends heavily on political support and institutional commitment that can vary across different government administrations. Building sustainable protection for indigenous technology rights requires ongoing organizing work and institutional development that goes beyond formal legal protections.

Regulatory Gaps and Systemic Limitations

Across different regulatory contexts, technology governance frameworks consistently exhibit democratic deficits that limit community participation while privileging technical expertise and corporate interests. These deficits reflect deeper tensions within liberal democratic systems between efficiency-oriented governance and meaningful democratic participation.

Regulatory processes typically treat technology governance as a technical problem requiring expert knowledge rather than a political challenge demanding democratic deliberation and community control. This approach systematically excludes community voices while legitimizing decisions that may impose significant costs on marginalized populations.

The complexity of technology systems creates barriers to community participation that regulatory frameworks rarely address through capacity building or institutional support. Communities affected by technology infrastructure often lack the technical expertise, legal resources, and time necessary for meaningful engagement with regulatory processes designed by and for technical experts.

Expert-dominated governance structures tend to reproduce existing power relationships while creating the appearance of inclusive decision-making. Even when regulatory frameworks include consultation requirements, these processes often function more as legitimation exercises than genuine opportunities for community influence over technology governance decisions.

Environmental Justice Integration Failures

Most technology governance frameworks fail to adequately integrate environmental justice principles that would prioritize community health and environmental protection over corporate profits or technical efficiency. This failure reflects the historical separation between technology policy and environmental regulation that treats these domains as distinct rather than interconnected.

Cumulative impact assessment approaches that consider how technology infrastructure contributes to existing environmental burdens remain rare in technology governance frameworks. This gap enables harmful infrastructure to concentrate in already overburdened communities while avoiding regulatory scrutiny that might prevent or mitigate community impacts.

Environmental justice communities often lack the resources and technical capacity necessary for meaningful participation in technology governance processes, while regulatory frameworks rarely provide the support and institutional accommodation necessary for inclusive participation. This dynamic reproduces environmental racism through seemingly neutral regulatory procedures.

Climate change considerations receive minimal attention in most technology governance frameworks despite the substantial energy consumption and carbon emissions associated with digital infrastructure. This gap reflects broader failures to integrate climate policy with other governance domains while treating environmental impacts as externalities rather than central considerations.

Global Governance Coordination Challenges

Technology systems operate across national boundaries while governance frameworks remain primarily national in scope, creating coordination challenges that enable companies to exploit regulatory arbitrage while limiting the effectiveness of protective regulations. This mismatch becomes particularly problematic for addressing global issues like climate change and international economic inequality.

International coordination mechanisms for technology governance remain weak and largely voluntary, enabling companies to resist binding regulations while maintaining the appearance of responsible governance through voluntary commitments and self-regulation initiatives. These voluntary approaches consistently prioritize corporate interests over community needs and environmental protection.

Trade agreements and international economic frameworks often constrain national technology governance capabilities by treating regulations as potential barriers to trade rather than legitimate expressions of democratic sovereignty. These constraints particularly affect developing countries that may have less bargaining power in international negotiations.

Global South countries face particular challenges in technology governance due to limited regulatory capacity, resource constraints, and economic pressures that may compromise environmental and social protections. Addressing these challenges requires international cooperation and resource sharing that goes beyond current aid and development frameworks.

Toward Democratic Technology Governance

The analysis of regulatory responses across different contexts reveals both the necessity and possibility of more democratic approaches to technology governance that center community needs, environmental protection, and global justice. Current regulatory frameworks, while offering important protections and accountability mechanisms, remain fundamentally inadequate for addressing the systemic inequalities and environmental harms produced by corporate-controlled technology development.

The most promising regulatory innovations examined in this chapter—from Nordic environmental assessment processes to Latin American community networking frameworks—demonstrate how governance systems can better serve community needs while maintaining technological capacity and environmental protection. However, these innovations remain constrained by broader political and economic systems that privilege corporate interests over democratic participation and environmental sustainability.

Building more democratic technology governance requires addressing several interconnected challenges: expanding community participation beyond consultation to meaningful decision-making power, integrating environmental justice principles into technology policy frameworks, developing international cooperation mechanisms that support rather than constrain democratic governance, and building community capacity for sustained engagement with complex technology governance processes.

The community organizing and resistance strategies examined in the previous chapter provide essential foundations for transformation, but sustainable change requires institutional innovations that can accommodate community demands while operating within existing political and economic constraints. This may require hybrid approaches that combine regulatory reform with community empowerment and alternative development models.

The urgency of climate change and the rapid pace of AI development make these governance challenges increasingly pressing. The next chapter will examine how international cooperation and policy innovation can support more democratic and environmentally sustainable approaches to technology development while addressing the global dimensions of technology justice.

The choice between corporate-controlled technology governance that serves profit over people and community-controlled governance that serves human needs and environmental protection remains ours to make. However, the window for effective action continues to narrow as corporate concentration increases and climate change accelerates. The regulatory frameworks examined in this chapter provide important tools for resistance and reform, but they cannot substitute for the broader social movements and alternative development models necessary for just and sustainable technology futures.

CHAPTER 12: TECHNICAL SOLUTIONS AND ACCOUNTABILITY

―――――――――――

Democratizing Technology Through Community-Controlled Transparency

When residents of South Memphis began investigating xAI's proposed data center in 2024, they faced a familiar challenge: how could communities without extensive technical training understand and challenge complex technological systems that would fundamentally alter their neighborhoods? The company's environmental impact statements were filled with technical jargon about "cooling efficiency metrics" and "power utilization effectiveness ratios" that obscured the basic questions residents needed answered: Would this facility poison their air? How much additional strain would it place on an already overburdened electrical grid? Who would profit from this development, and who would bear its costs?

The Memphis organizing campaign succeeded precisely because community researchers developed accessible methods for translating technical complexity into community knowledge while building analytical capacity that could challenge corporate expertise on its own terms. Working with environmental scientists, data analysts, and community health experts, residents created parallel knowledge systems that could evaluate corporate claims while remaining grounded in community experience and values. This work exemplified emerging approaches to technical accountability that treat transparency not as a technical problem requiring expert solutions, but as a political challenge demanding community empowerment and democratic control over knowledge production.

The development of community-controlled technical accountability mechanisms represents a crucial frontier in the struggle for technology justice. While previous chapters have examined regulatory frameworks and resistance strategies, this chapter focuses on how affected communities can develop the analytical tools necessary for understanding, challenging, and governing complex technological systems. These approaches recognize that technical transparency without community power remains meaningless, while community organizing without technical capacity may prove insufficient for challenging sophisticated systems of algorithmic oppression.

Explainable AI: Beyond Technical Compliance

The Limits of Corporate Explainability

The technology industry's embrace of "explainable AI"(XAI) represents both an acknowledgment that algorithmic systems require transparency and a sophisticated attempt to define transparency in ways that serve corporate interests rather than community needs. Current XAI approaches typically focus on making algorithmic decision-making processes interpretable to technical experts and regulatory authorities while providing minimal tools for community understanding or democratic oversight.

Most explainability frameworks treat transparency as a technical property of algorithmic systems rather than a social relationship between communities and the institutions governing their lives. This approach produces explanations optimized for technical accuracy rather than community comprehension, using statistical measures and mathematical concepts that may be meaningless to people experiencing algorithmic harm but lacking formal training in data science or machine learning.

Corporate XAI implementations often focus on post-hoc explanations that describe how algorithmic systems reached particular decisions without revealing the underlying assumptions, data sources, or value judgments embedded in system design. These explanations can actually obscure accountability by suggesting that algorithmic decisions result from neutral technical processes rather than political choices about whose interests algorithmic systems should serve.

The emphasis on individual explanations for specific algorithmic decisions reflects broader patterns within liberal governance that treat systemic problems as collections of individual cases rather than addressing structural inequalities that produce discriminatory outcomes. This individualistic approach enables companies to provide explanations for particular decisions while avoiding accountability for systematic patterns of discrimination or community harm.

Community-Centered Explainability Frameworks

Developing explainable AI systems that serve community needs requires fundamentally different approaches that prioritize collective understanding over individual explanations while treating transparency as a foundation for democratic governance rather than an end in itself. Community-centered explainability begins with questions that communities need answered rather than technical metrics that systems can easily provide.

Effective community explainability frameworks address several key questions: How do algorithmic systems affect community wellbeing and environmental health? What assumptions about community needs and values are embedded in system design? How do algorithmic decisions interact with other institutional systems affecting community life? What alternatives to current algorithmic approaches might better serve community interests? Who profits from current system configurations, and how might communities capture more benefits from technological development?

These questions require explanatory approaches that connect algorithmic system performance to broader patterns of community impact while providing communities with information necessary for informed participation in technology governance decisions. This may involve translating technical performance metrics into community-relevant outcomes, documenting cumulative impacts across multiple technological systems, and revealing connections between algorithmic decisions and other institutional practices affecting community life.

Community-centered explainability must also address power imbalances that prevent communities from accessing and interpreting technical information. This requires not only simplified explanations but also capacity-building approaches that enable communities to develop independent analytical capabilities while maintaining democratic control over knowledge production processes.

Participatory Algorithm Design and Community Input

Moving beyond post-hoc explainability requires integrating community participation into algorithmic system design processes from initial conception through ongoing operation and evaluation. Participatory design approaches recognize communities as co-creators rather than passive recipients of technological systems while ensuring that community values and priorities shape algorithmic development.

Participatory algorithm design typically involves several stages: community problem definition that identifies issues requiring algorithmic intervention based on community experience rather than technical assumptions, collaborative system design that involves community members in defining algorithmic approaches and success metrics, iterative development that enables ongoing community input on system performance and modification, and community-controlled evaluation that assesses algorithmic impact according to community-defined criteria.

These processes require significant time, resources, and institutional support that may exceed the capacity of many community organizations. Successful participatory design efforts often involve partnerships between community organizations and sympathetic technical experts who can provide analytical support while respecting community leadership and democratic decision-making processes.

However, participatory design approaches face ongoing challenges from institutional pressures that prioritize technical efficiency and economic optimization over community participation and democratic governance. Balancing community input with technical constraints requires careful negotiation and may produce algorithmic systems that perform differently than those optimized purely for technical metrics.

Cultural Competence in Technical Explanation

Developing explainable AI systems that serve diverse communities requires cultural competence that recognizes how different communities may understand and evaluate technological systems according to distinct values, knowledge traditions, and communication practices. This goes beyond language translation to encompass fundamentally different approaches to knowledge, evidence, and decision-making.

Indigenous communities, for example, may evaluate algorithmic systems according to principles of intergenerational responsibility, ecological sustainability, and collective decision-making that differ significantly from dominant Western approaches emphasizing individual rights, economic efficiency, and technical optimization. Explanatory frameworks must accommodate these different evaluation criteria rather than imposing uniform approaches across diverse communities.

Similarly, communities with different educational backgrounds, economic resources, or relationships to formal institutions may require different explanatory approaches that build on existing knowledge systems rather than requiring communities to adopt technical frameworks developed for expert audiences. This may involve using community storytelling traditions, visual communication methods, or experiential learning approaches that connect algorithmic concepts to community experience.

Cultural competence also requires recognizing how algorithmic systems may interact with community governance traditions, conflict resolution practices, and collective decision-making processes. Explanatory frameworks should support rather than undermine these community institutions while providing information necessary for communities to maintain control over technological systems affecting their lives.

Community-Based Algorithmic Auditing

Grassroots Technical Investigation Methods

Community-based algorithmic auditing represents a fundamental shift from expert-dominated accountability mechanisms toward democratic oversight processes that center affected communities as primary evaluators of algorithmic system performance. These approaches recognize that communities experiencing algorithmic harm possess essential knowledge about system impacts that technical experts may overlook or minimize.

Community auditing methods typically combine technical analysis with ethnographic investigation, policy research, and organizing strategies to produce comprehensive assessments of algorithmic system impacts on community life. This interdisciplinary approach enables communities to evaluate not only technical performance metrics but also broader questions about algorithmic system integration with community institutions, cultural practices, and social relationships.

The Memphis xAI investigation exemplifies community auditing approaches by combining environmental monitoring, energy grid analysis, economic impact assessment, and community health evaluation to produce a comprehensive understanding of data center impacts that corporate environmental assessments systematically ignore. Community researchers documented how the facility would interact with existing pollution sources while examining economic arrangements that would privatize profits while socializing environmental and health costs.

Community auditing approaches often reveal algorithmic system impacts that formal evaluation methods miss because they occur at scales, timelines, or social locations that technical assessments do not examine. For example, community investigators may document how algorithmic hiring systems affect neighborhood economic stability, how predictive policing algorithms influence community social relationships, or how automated benefit determination systems interact with existing poverty and discrimination.

Collaborative Research Partnerships

Effective community-based algorithmic auditing often requires partnerships between community organizations and sympathetic researchers who can provide technical expertise while respecting community leadership and democratic decision-making processes. These partnerships must navigate power imbalances that could reproduce expert domination while enabling communities to access technical resources necessary for sophisticated algorithmic analysis.

Successful community-researcher partnerships typically involve several key principles: community control over research questions and priorities, transparent resource sharing that ensures community organizations receive appropriate compensation for their contributions, capacity building that enables community members to develop independent analytical capabilities, and shared ownership of research products that ensures communities maintain control over how findings are used.

The Community-Based Participatory Research (CBPR) tradition provides important models for equitable research partnerships that center community knowledge while incorporating technical expertise. CBPR approaches recognize community members as co-researchers capable of identifying research questions, collecting data, analyzing findings, and developing policy recommendations based on research results.

However, algorithmic auditing partnerships face unique challenges due to the technical complexity of contemporary AI systems and the resources required for sophisticated technical analysis. Community organizations may lack the funding necessary to support sustained technical investigation while researchers may face institutional pressures that conflict with community priorities and democratic decision-making processes.

Digital Rights and Privacy Protection

Community-based algorithmic auditing must address significant privacy and digital rights challenges that arise when communities investigate algorithmic systems affecting their lives. Many algorithmic systems operate using proprietary data and algorithms that companies protect as trade secrets, creating legal and technical barriers to community investigation.

Community auditors have developed innovative approaches to these challenges, including reverse engineering techniques that infer algorithmic behavior from observable outputs, freedom of information requests that access government algorithmic systems and data, collaborative data collection that enables communities to pool information about algorithmic impacts, and legal advocacy that challenges trade secrecy claims when they prevent accountability for systems affecting public welfare.

These approaches require careful attention to privacy protection for community members while maintaining transparency about algorithmic system impacts. Community auditing projects must balance the need for detailed information about algorithmic harm with requirements to protect individual privacy and prevent retaliation against community members participating in investigation efforts.

Digital security considerations become particularly important when community auditing efforts challenge powerful corporations or government agencies that may attempt to prevent or undermine community investigation. Community organizations need access to secure communication tools, data protection technologies, and legal support that enables sustained investigation while protecting community members from surveillance or harassment.

Documentation and Evidence Standards

Community-based algorithmic auditing requires developing documentation and evidence standards that can support both community organizing and formal accountability processes while remaining accessible to community members without technical training. These standards must balance rigor with accessibility while producing evidence that can withstand scrutiny from corporate legal teams and technical experts.

Effective community documentation approaches typically combine quantitative analysis with qualitative investigation, policy research, and community testimony to produce comprehensive records of algorithmic system impacts. This multi-method approach enables communities to document technical performance while also capturing social, cultural, and economic impacts that quantitative methods may miss.

Community documentation must also address temporal and spatial scales that formal evaluation methods often ignore. Algorithmic systems may produce cumulative impacts over years or decades while affecting geographic areas larger than individual communities. Community auditing approaches need to develop methods for tracking these long-term and large-scale impacts while maintaining focus on specific community experiences.

Quality control and peer review processes adapted to community contexts can help ensure that community investigations meet standards necessary for public credibility while maintaining democratic participation in knowledge production. This may involve partnerships with academic institutions, collaboration between multiple community organizations, or development of community-controlled review processes that evaluate investigation methods and findings.

Digital Forensic Analysis for Discriminatory Systems

Developing Community Forensic Capacity

Digital forensic analysis represents a crucial tool for exposing discriminatory algorithmic systems and challenging corporate claims about algorithmic neutrality and fairness. However, traditional digital forensics focuses on law enforcement applications rather than community empowerment and democratic accountability. Developing community-controlled forensic capabilities requires adapting technical methods while building institutional capacity that serves community organizing rather than state surveillance.

Community digital forensics typically involves several interconnected activities: documenting algorithmic decision-making processes and their impacts on community members, analyzing algorithmic system performance across different demographic groups and geographic areas, investigating algorithmic system integration with other institutional practices and discriminatory systems, and developing evidence that can support community organizing, legal advocacy, and policy change efforts.

The Memphis data center investigation demonstrates how communities can develop forensic analysis capabilities by combining technical investigation with policy research and environmental monitoring. Community researchers used publicly available data about energy consumption, environmental permits, and corporate financial arrangements to document how xAI's development plans would systematically extract resources from Black communities while privatizing profits to wealthy investors.

This forensic approach revealed discriminatory patterns that corporate environmental assessments systematically obscured: the company specifically targeted a predominantly Black neighborhood already bearing disproportionate environmental burdens while avoiding wealthier, whiter areas with similar technical characteristics. By analyzing site selection criteria alongside demographic data and environmental health outcomes, community researchers documented environmental racism that violated both legal requirements and community values.

Reverse Engineering Discriminatory Algorithms

Reverse engineering techniques enable communities to understand how algorithmic systems operate when companies refuse to provide transparency or when corporate explanations prove inadequate for community accountability needs. These techniques involve analyzing algorithmic system inputs and outputs to infer decision-making processes while documenting discriminatory impacts that formal evaluation methods may miss.

Reverse engineering approaches for community accountability typically focus on questions that matter for community organizing rather than technical optimization: How do algorithmic systems treat different demographic groups or geographic areas? What assumptions about community needs and values are embedded in algorithmic design? How do algorithmic decisions interact with other institutional practices affecting community life? What evidence exists of discriminatory impact or intent?

Community reverse engineering efforts often reveal discriminatory patterns that companies deny or minimize in public communications. For example, community investigators have documented how hiring algorithms systematically disadvantage applicants from certain neighborhoods, how credit scoring systems perpetuate racial wealth gaps, and how predictive policing algorithms concentrate surveillance in communities of color re-gardless of actual crime patterns.

These investigations require significant technical sophistica-tion and resources that may exceed the capacity of individual community organizations. However, collaborative approaches that pool resources across multiple communities and organiza-tions can enable more comprehensive analysis while building shared analytical capacity that serves broader movement goals.

Exposing Environmental Racism in Technology Infrastructure

Digital forensic analysis proves particularly valuable for ex-posing environmental racism in technology infrastructure de-ployment because discriminatory patterns often remain hidden within technical decision-making processes that appear race-neutral while producing racially disparate outcomes. Community forensic investigation can reveal how seemingly technical choices about site selection, environmental impact assessment, and com-munity consultation reproduce historical patterns of environ-mental injustice.

The process of exposing environmental racism typically in-volves several analytical steps: spatial analysis that maps tech-nology infrastructure deployment patterns against demographic characteristics and existing environmental burdens, historical in-vestigation that documents how current discriminatory patterns connect to previous exclusionary practices, institutional analysis that examines decision-making processes for evidence of dis-criminatory intent or impact, and cumulative impact assessment that considers how technology infrastructure adds to communi-ties' existing environmental burdens.

Community forensic teams investigating the Memphis xAI fa-cility documented how the company's site selection process systematically avoided white neighborhoods while targeting a predominantly Black area already experiencing disproportionate pollution and health impacts. This analysis revealed that technical factors like electrical grid capacity and transportation access—which the company cited as determinative—were actually present in multiple locations, but the company chose the site that would impose environmental costs on the most vulnerable population.

The investigation also documented how corporate environmental assessments systematically minimized community health impacts by using outdated baseline data, failing to consider cumulative impacts from multiple pollution sources, and excluding community health priorities from impact evaluation criteria. This analysis enabled community organizers to challenge both specific permit applications and broader environmental review processes that enable environmental racism.

Building Legal and Policy Evidence

Digital forensic analysis for community accountability must produce evidence that can support legal challenges, policy advocacy, and regulatory enforcement while remaining grounded in community experience and values. This requires developing analytical approaches that meet legal evidence standards while maintaining accessibility for community organizing and public education efforts.

Legal applications of community digital forensics typically focus on proving discriminatory intent or impact in violation of civil rights laws, environmental regulations, or administrative procedures. This requires documentation that demonstrates how algorithmic systems or infrastructure deployment decisions systematically disadvantage protected groups while showing that these patterns result from institutional choices rather than neutral technical factors.

Community forensic evidence has proven effective in challenging discriminatory algorithmic systems through Title VI civil rights complaints, environmental justice lawsuits, and administrative appeals of government decisions. However, these legal strategies require significant resources and expertise that may exceed community capacity while involving long timelines that may not align with urgent community needs.

Policy advocacy applications may prove more accessible for many community organizations because they require less formal legal evidence while enabling communities to influence regulatory processes, legislative developments, and public opinion. Community forensic analysis can support policy advocacy by documenting system-wide discriminatory patterns, revealing gaps in current oversight mechanisms, and demonstrating community capacity for independent analysis and oversight.

Making Technical Systems Transparent to Affected Communities

Community Technology Education and Literacy

Making technical systems transparent to affected communities requires educational approaches that build community analytical capacity while respecting existing knowledge systems and cultural practices. This goes beyond traditional technology literacy to encompass critical analysis of how technological systems interact with community life while developing skills necessary for democratic participation in technology governance.

Community technology education must address several interconnected goals: developing technical understanding sufficient for meaningful participation in technology governance decisions, building critical analytical capacity for evaluating corporate and government claims about technological systems, creating collaborative learning processes that strengthen community solidarity while building individual skills, and connecting technology analysis to broader community organizing and social justice work.

Effective community technology education typically combines formal educational components with experiential learning opportunities that enable community members to investigate technological systems affecting their own lives. This approach recognizes that people learn most effectively when education connects to immediate concerns and provides tools for addressing problems they already face.

The Memphis organizing campaign demonstrates how community education can develop through collective investigation and action rather than formal classroom instruction. As residents worked together to understand xAI's development plans, they developed shared analytical frameworks for evaluating corporate claims while building technical capacity that could be applied to other community challenges.

Accessible Technical Communication

Translating complex technical information into accessible community communication requires approaches that respect community intelligence while acknowledging that technical complexity can create barriers to meaningful participation in technology governance. This involves more than simplification—it requires developing communication strategies that connect technical concepts to community experience while providing tools for independent analysis and verification.

Accessible technical communication typically involves several strategies: using concrete examples and analogies that connect technical concepts to familiar community experiences, providing multiple communication formats including visual, narrative, and

interactive approaches that accommodate different learning preferences, creating opportunities for questions and dialogue rather
than one-way information transmission, and connecting technical information to community values and organizing goals rather
than treating technical analysis as an end in itself.

Community-controlled communication approaches often
prove more effective than expert-led education because they
emerge from community priorities and use communication
methods that community members find engaging and useful.
Community members who develop technical understanding
through collective investigation often become effective educators
for other community members because they can translate technical concepts using shared cultural references and communication practices.

However, accessible communication must balance simplicity
with accuracy while avoiding technical explanations that mislead
community members about complex systems. This requires careful attention to which technical details matter for community
decision-making while ensuring that simplified explanations
provide adequate foundations for meaningful participation in
technology governance.

Democratic Oversight Mechanisms

Creating meaningful community oversight of technical systems requires institutional innovations that provide communities
with real decision-making power rather than merely consultation
opportunities. This involves developing governance mechanisms
that can accommodate technical complexity while maintaining
democratic accountability and community control over decisions
affecting community life.

Effective democratic oversight typically involves several components: community-controlled technical advisory capacity that
provides independent analysis of technological systems, regular

public reporting requirements that ensure community access to information about system performance and impacts, meaningful consultation processes that involve communities in ongoing system governance rather than one-time input opportunities, and accountability mechanisms that enable communities to modify or remove technological systems that fail to serve community needs.

Community oversight boards with real authority over technological systems represent one approach to democratic governance that enables sustained community engagement while providing institutional mechanisms for technical accountability. However, these boards require significant resources, technical support, and legal authority to function effectively as community governance institutions.

Alternative approaches might include community-controlled research partnerships that enable ongoing investigation of technological systems, participatory technology assessment processes that involve communities in evaluating emerging technologies, or community ownership models that give communities direct control over technological infrastructure affecting their lives.

Building Independent Analytical Infrastructure

Sustaining community transparency and accountability efforts requires building independent analytical infrastructure that enables communities to conduct ongoing investigation and oversight without depending on corporate or government cooperation. This involves developing technical capacity, institutional resources, and collaborative networks that support sustained community engagement with complex technological systems.

Independent analytical infrastructure typically includes several components: technical tools and resources that enable community investigation of algorithmic systems and technology infrastructure, training and capacity building programs that de-velop community analytical capabilities, funding and organiza-tional support that enables sustained investigation and oversight efforts, and collaborative networks that enable resource sharing and coordination between community organizations.

Building this infrastructure requires significant resources and long-term commitment that may challenge community organiza-tions already struggling to address urgent immediate needs. How-ever, communities that develop independent analytical capacity often find that these capabilities prove valuable for addressing multiple issues while building community power and organiza-tional capacity.

The development of community-controlled analytical infra-structure also faces resistance from corporations and government agencies that prefer to maintain control over technical infor-mation and analysis. Communities may need to combine trans-parency advocacy with legal strategies, direct action, and coalition building to secure access to information and resources necessary for independent oversight.

Integrating Technical Accountability with Community Organizing

Technical accountability mechanisms prove most effective when they connect directly to community organizing strategies and political goals rather than treating transparency as an end in itself. This requires developing analytical approaches that pro-vide communities with tools for challenging harmful technologi-cal systems while building power for broader social and economic transformation.

Effective integration of technical analysis with community or-ganizing typically involves several principles: starting with com-munity priorities and organizing goals rather than technical questions, developing analysis that supports community action strategies while building shared understanding of technological systems, creating opportunities for community members to participate directly in technical investigation rather than receiving expert analysis passively, and connecting specific technological issues to broader patterns of oppression and resistance.

The Memphis organizing campaign exemplifies this integrated approach by combining technical analysis of environmental impacts with broader organizing around economic development, racial justice, and community self-determination. Community researchers used technical investigation to support permit challenges while also building community capacity for ongoing engagement with development decisions affecting their neighborhood.

This approach enables communities to use technical analysis strategically while maintaining focus on broader organizing goals rather than becoming absorbed in technical details that may distract from political action. Technical investigation becomes a tool for community empowerment rather than an alternative to organizing or political engagement.

Scaling Community Technical Capacity

Individual community organizations often lack the resources necessary for sophisticated technical analysis, but collaborative approaches can enable communities to pool resources while building shared analytical capacity that serves broader movement goals. This requires developing institutional innovations that support resource sharing while maintaining community control over analytical priorities and methods.

Regional networks of community organizations can share technical resources while maintaining local autonomy over how analytical capabilities are used. Examples include collaborative hiring of technical staff, shared funding for expensive analytical tools, coordinated training programs that build capacity across multiple organizations, and joint investigation projects that address shared concerns while building analytical capabilities.

Academic partnerships can provide technical resources and expertise while supporting community priorities rather than imposing academic research agendas. However, these partnerships require careful attention to power dynamics that could reproduce expert domination while ensuring that communities maintain control over research questions, methods, and findings.

Movement-wide analytical infrastructure might include shared technical tools, collaborative training programs, coordinated funding strategies, and policy advocacy that supports community analytical capacity. Building this infrastructure requires long-term commitment and sustained cooperation between community organizations with different priorities and approaches.

Policy and Legal Integration

Community technical analysis proves most effective when it connects to broader policy advocacy and legal strategies rather than remaining isolated within individual community campaigns. This requires developing analytical approaches that can support multiple forms of political engagement while maintaining community control over how technical findings are used.

Policy applications of community technical analysis include supporting legislative advocacy for stronger oversight requirements, providing evidence for regulatory enforcement actions, contributing to public comment processes on government technology policies, and documenting community impacts for policy development purposes.

Legal applications might involve supporting civil rights complaints about discriminatory algorithmic systems, challenging environmental permits for harmful technology infrastructure, appealing government decisions about technology deployment, or defending community members facing harm from technological systems.

However, legal and policy strategies can consume significant resources while involving timelines and procedures that may not align with community organizing priorities. Communities need to balance engagement with formal political processes with sustained organizing work that builds community power independently of institutional recognition or support.

Toward Community-Controlled Technology

The technical accountability mechanisms examined in this chapter represent essential tools for challenging harmful algorithmic systems and building community power over technological development. However, these tools prove most effective when integrated with broader organizing strategies that address the systemic inequalities enabling technological oppression rather than focusing solely on improving individual technological systems.

Community-controlled explainable AI, algorithmic auditing, digital forensics, and transparency mechanisms provide foundations for more democratic approaches to technology governance, but they cannot substitute for the economic redistribution, institutional transformation, and cultural change necessary for technology justice. These technical tools must support rather than replace community organizing, policy advocacy, and alternative development strategies examined in previous chapters.

The Memphis organizing campaign demonstrates how communities can successfully integrate technical analysis with broader organizing to challenge specific harmful projects while building capacity for ongoing resistance to digital colonialism.

Community researchers combined environmental monitoring, economic analysis, and policy investigation to document how xAI's development plans would systematically harm their neighborhood while building community understanding and analytical capacity for addressing future challenges.

However, individual victories against harmful technology projects cannot substitute for broader transformation of the systems producing technological injustice. The next chapter will examine how climate justice considerations and environmental sustainability requirements can reshape technology development while supporting rather than undermining community control and democratic governance.

The development of community-controlled technical accountability represents a crucial step toward more democratic technology futures, but it requires sustained commitment and resources that may challenge community organizations already struggling with immediate needs. Building the analytical infrastructure necessary for meaningful community oversight of complex technological systems requires long-term vision and collaborative approaches that connect individual community campaigns to broader movements for social and economic justice.

The choice between expert-dominated technical governance that serves corporate interests and community-controlled accountability that serves human needs remains ours to make. The tools and strategies examined in this chapter provide communities with concrete methods for challenging technological oppression while building power for broader transformation. However, their effectiveness depends on sustained community organizing and political engagement that can maintain democratic pressure for accountability while building alternative approaches to technological development that serve community needs rather than corporate profits.

PART VI: FUTURE DIRECTIONS

CHAPTER 13: CLIMATE JUSTICE & SUSTAINABLE TECHNOLO

The Planetary Scale of Algorithmic Environmental Injustice

The artificial intelligence industry's explosive growth has created an unprecedented demand for computational infrastructure that reveals fundamental contradictions within the digital economy. While tech companies promote AI as a clean, ethereal technology that operates in virtual "clouds," the material reality involves massive data centers, energy-intensive processing facilities, and complex supply chains that systematically concentrate environmental burdens in communities of color worldwide.

Contemporary AI development operates through sophisticated mechanisms of spatial displacement that enable wealthy technology centers to maintain pristine corporate environments while externalizing environmental costs to vulnerable populations. Silicon Valley companies power their gleaming headquarters through coal plants in Appalachian communities, hydroelectric installations that displace indigenous populations, and rare earth mining operations that contaminate villages across the Global South. This arrangement represents more than corporate environmental irresponsibility—it embodies a global system of technological environmental racism that reproduces colonial patterns of resource extraction while claiming technological neutrality.

The environmental footprint of AI extends far beyond energy consumption to encompass water usage for cooling systems, electronic waste from rapidly obsolete hardware, and toxic materials required for semiconductor production. Data centers processing machine learning algorithms consume water at rates comparable to small cities, often drawing from aquifers serving predominantly Black and Hispanic communities already struggling with environmental contamination. Meanwhile, the rare earth elements essential for AI hardware come primarily from mines in Africa and Asia where environmental and health impacts fall disproportionately on indigenous and minority populations.

The planetary scale of AI infrastructure demands analysis that moves beyond individual cases of environmental harm to examine the global systems enabling technological development that serves corporate profits while imposing environmental costs on marginalized communities worldwide. This chapter examines how climate change, energy scarcity, and environmental degradation intersect with algorithmic development to create new forms of environmental racism that operate across national boundaries while maintaining the fiction of clean technology through careful geographic separation of benefits and burdens.

The Impossible Geography of AI Infrastructure

Energy Consumption and Physical Limits.

The exponential growth in AI computational requirements has created energy demands that increasingly exceed the physical and environmental limits of sustainable development. Training a single large language model can consume energy equivalent to the lifetime emissions of multiple automobiles, while the global AI infrastructure already accounts for approximately 4% of global greenhouse gas emissions—a figure projected to reach 8-14% by 2030 if current growth trajectories continue.

These energy requirements reflect fundamental physical constraints that no amount of technological innovation can eliminate. The thermodynamic limits of computation mean that processing information requires energy expenditure that increases with computational complexity, while the speed-of-light limitations on data transmission necessitate physical infrastructure that cannot be virtualized or dematerialized regardless of software advances.

The exponential growth in AI energy consumption represents more than a technical challenge—it reveals the fundamental unsustainability of current technological development trajectories that systematically impose environmental costs on marginalized communities while concentrating benefits among wealthy elites. Current projections indicate that AI systems will consume between 4-14% of global electricity by 2030, a trajectory that is physically incompatible with climate stabilization goals while following predictable patterns of environmental racism in infrastructure deployment. The following analysis demonstrates how this energy crisis intersects with digital colonialism, as the massive computational requirements driving energy demand are met through strategic extraction from communities with limited political capacity to resist environmental burden.

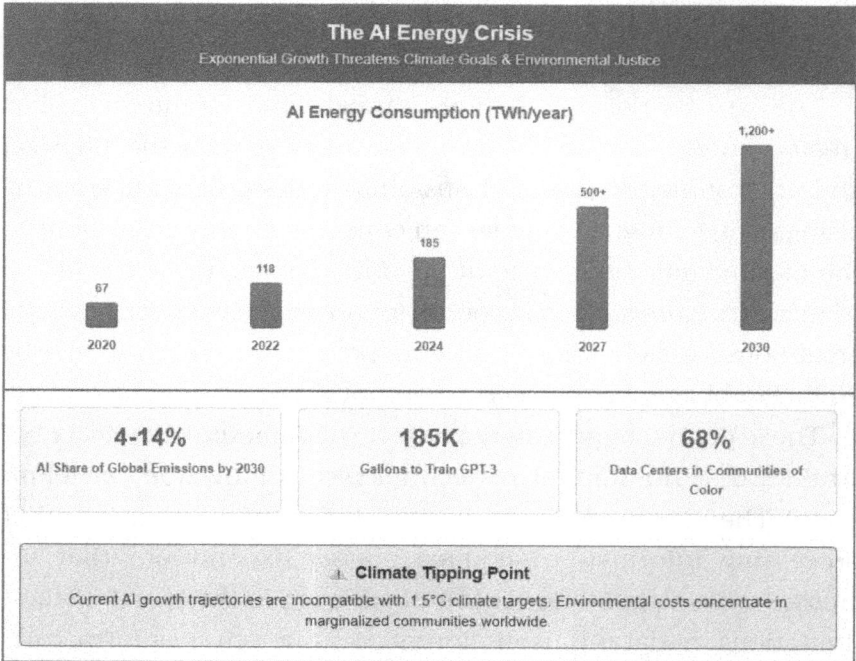

The AI Energy Crisis
Exponential Growth Threatens Climate Goals & Environmental Justice

AI Energy Consumption (TWh/year)

67	118	185	500+	1,200+
2020	2022	2024	2027	2030

4-14%	185K	68%
AI Share of Global Emissions by 2030	Gallons to Train GPT-3	Data Centers in Communities of Color

⚠ Climate Tipping Point
Current AI growth trajectories are incompatible with 1.5°C climate targets. Environmental costs concentrate in marginalized communities worldwide.

Current AI development trajectories assume unlimited energy availability and environmental absorption capacity that simply do not exist within planetary boundaries. The industry's pursuit of artificial general intelligence and beyond requires computational resources that would consume substantial portions of global energy production while generating carbon emissions incompatible with climate stabilization goals.

The geographic distribution of this energy consumption reveals systematic patterns of environmental racism that concentrate clean technology development in wealthy regions while externalizing energy production and environmental degradation to marginalized communities. Silicon Valley's gleaming corporate campuses operate through supply chains extending to coal-powered data centers in rural America, hydroelectric facilities displacing indigenous communities in the Global South, and manufacturing facilities polluting working-class communities worldwide.

Continental Overflow and Infrastructure Extraction

The concept of "continental overflow" describes how AI development in wealthy regions increasingly depends on infrastructure extraction from other continents to sustain energy-intensive computational requirements that exceed local environmental capacity. This process operates through sophisticated financial and technical arrangements that enable technology companies to access global energy resources while maintaining legal and political distance from environmental consequences.

The xAI power plant import exemplifies this dynamic by demonstrating how American AI companies increasingly rely on infrastructure developed elsewhere to meet computational demands that cannot be sustained within existing domestic energy systems. Rather than reducing computational requirements or accepting energy limitations, companies pursue technological solutions that displace environmental costs to other regions while maintaining developmental autonomy.

This infrastructure extraction reproduces colonial patterns of resource appropriation while creating new forms of technological dependence that constrain sovereignty in both exporting and importing regions. Countries providing energy infrastructure may find their environmental capacity absorbed by foreign computational requirements while importing regions become dependent on continued energy extraction that may prove environmentally or politically unsustainable.

The financial arrangements enabling infrastructure extraction typically involve complex corporate structures, international trade agreements, and development finance mechanisms that ob-

scure environmental costs while concentrating economic benefits among wealthy investors. These arrangements enable
companies to externalize environmental risks while privatizing
technological benefits, creating incentive structures that systematically favor environmental degradation over sustainable development.

Technical Impossibilities and System Vulnerabilities

Current AI development trajectories face fundamental technical constraints that cannot be overcome through incremental
efficiency improvements or technological innovation alone. The
energy requirements for training increasingly sophisticated models grow exponentially while efficiency gains proceed linearly,
creating an unsustainable trajectory that will eventually collide
with physical and environmental limits.

The semiconductor manufacturing processes underlying AI
hardware require enormous energy inputs, toxic chemical
processes, and rare earth materials with devastating environmental consequences. Manufacturing a single advanced processor can
require thousands of gallons of ultrapure water, dozens of toxic
chemicals, and energy equivalent to operating a household for
months—environmental costs that multiply across the millions
of processors required for global AI infrastructure.

Data center cooling requirements present additional physical
constraints that cannot be eliminated through software improvements. High-performance computing generates enormous heat
that must be dissipated through energy-intensive cooling systems, creating secondary energy demands that often exceed the
computational energy requirements themselves.

The geographic constraints on data center placement reflect
these physical realities: facilities require enormous quantities of

cooling water, stable electrical grid connections, and environmental conditions that can accommodate industrial heat generation. These requirements systematically exclude many regions while concentrating infrastructure in areas with abundant water resources and electrical capacity—often coinciding with marginalized communities lacking political power to resist harmful development.

Global Environmental Case Studies: Tech Infrastructure in the Global South

Rare Earth Mining and Electronic Waste in Africa

The Democratic Republic of Congo produces over 60% of the world's cobalt, an essential component in the lithium-ion batteries powering global AI infrastructure. Mining operations in the DRC operate through brutal labor practices, environmental devastation, and political violence that enable technology companies to access essential materials while maintaining legal distance from human rights violations and environmental destruction.

Cobalt mining in the DRC involves child labor, unsafe working conditions, and environmental contamination that poisons local water supplies while generating profits for international technology companies. Mining operations displace rural communities, destroy agricultural land, and create long-term environmental damage that persists decades after extraction concludes.

The contrast between cobalt mining conditions in the DRC and Silicon Valley's clean corporate environments reveals the global geography of technological environmental racism: essential materials are extracted through environmentally devastating processes in African communities while finished products generate profits for predominantly white technology executives and investors in American urban centers.

The global deployment of AI infrastructure follows systematic patterns that concentrate environmental burdens in communities least equipped to resist corporate extraction while directing technological benefits to wealthy regions. Comprehensive analysis of data center placement worldwide reveals that 68% of facilities are strategically located in communities of color, despite these populations representing only 13% of the U.S. population. This spatial targeting represents a calculated strategy that exploits existing environmental burdens and limited political representation to impose additional harm through what can only be described as algorithmic environmental colonialism. The following analysis documents how this pattern extends internationally, creating sacrifice zones from Chile to India to Ghana.

Global Pattern: Data Center Environmental Burden Distribution

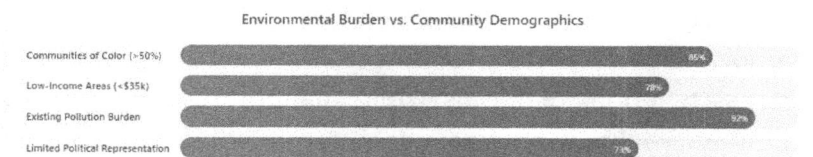

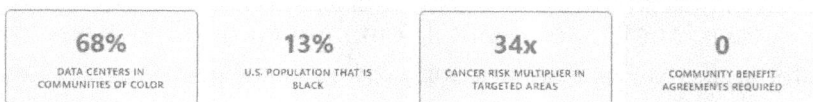

Environmental Burden vs. Community Demographics

Communities of Color (>50%)	86%
Low-Income Areas (<$35k)	78%
Existing Pollution Burden	92%
Limited Political Representation	74%

68%	13%	34x	0
DATA CENTERS IN COMMUNITIES OF COLOR	U.S. POPULATION THAT IS BLACK	CANCER RISK MULTIPLIER IN TARGETED AREAS	COMMUNITY BENEFIT AGREEMENTS REQUIRED

International Environmental Colonialism

Quilicura, Chile (Google)		Bengaluru, India (Multiple)		Agbogbloshie, Ghana (E-waste)	
Water Extraction	1B+ liters/year	Daily Water Consumption	8M liters	E-waste Processing	Toxic informal sector
Local Water Access	Severe drought	Recent Water Crisis	Worst in 500 years	Health Impacts	Severe lead/mercury exposure
Primary Beneficiaries	Global North users	Community Consultation	Minimal	Corporate Accountability	None

Global Environmental Colonialism Pattern

Tech companies systematically export environmental costs to Global South communities while concentrating digital benefits in wealthy regions. This represents a new form of colonial extraction that uses algorithmic systems to obscure traditional patterns of resource appropriation.

Ghana's Agbogbloshie electronic waste dump processes discarded technology from wealthy countries through informal recycling operations that expose workers to toxic materials while recovering valuable metals for global supply chains. Workers, predominantly young men from marginalized communities, burn electronic waste to extract copper and other valuable materials, creating toxic smoke that pollutes local air and water while generating severe health consequences.

The electronic waste trade operates through international agreements and corporate practices that enable wealthy countries to export environmental costs while importing processed materials for new technology production. This arrangement creates circular systems of environmental racism that concentrate technology benefits in wealthy regions while imposing both extraction and disposal costs on Global South communities.

Data Center Colonialism in Latin America

Technology companies increasingly establish data centers in Latin American countries to access cheaper energy, favorable regulatory environments, and geographic proximity to growing regional markets. However, these facilities often reproduce colonial patterns of resource extraction while providing minimal local benefits and imposing substantial environmental costs on host communities.

Costa Rica's data center development demonstrates both opportunities and risks associated with technology infrastructure in the Global South. The country's renewable energy resources enable relatively clean data center operations while government policies emphasize environmental protection and sustainable development. However, even renewable-powered facilities require enormous water consumption for cooling while generating electronic waste and contributing to regional energy demand that may strain electrical grid capacity.

Chile's hyperscale data center development in the Atacama Desert illustrates how companies exploit regulatory arbitrage and resource availability to minimize operational costs while externalizing environmental risks. The region's dry climate reduces cooling requirements while remote location minimizes political opposition, but water consumption for cooling competes with indigenous communities' traditional water rights while contributing to regional environmental stress.

Brazilian Amazon data center proposals reveal the extreme end of environmental colonialism in technology infrastructure placement. Companies cite the region's cooling potential and renewable energy availability while ignoring massive environmental risks associated with industrial development in ecologically sensitive areas. These proposals demonstrate how technology companies conceptualize the natural environment as a resource for computational optimization rather than a complex ecosystem requiring protection and stewardship.

Asian Manufacturing and Environmental Degradation

China's role as the global center for electronics manufacturing creates environmental conditions that enable worldwide technology consumption while concentrating pollution and health risks in Chinese working-class communities. Manufacturing facilities producing semiconductors, batteries, and electronic components generate air and water pollution that affects millions of people while enabling clean technology consumption in wealthy countries.

The city of Guiyu in Guangdong Province has become synonymous with electronic waste processing through informal recycling operations that expose workers and residents to toxic materials while recovering valuable components for global supply chains. Recycling operations involve acid baths, open burning, and primitive disassembly techniques that contaminate local air, water, and soil while generating severe health consequences for exposed populations.

Taiwan's semiconductor manufacturing industry demonstrates how technology production can generate enormous environmental costs even in wealthy regions with strong environmental regulations. TSMC, the world's largest contract semiconductor manufacturer, consumes massive quantities of water and energy while generating chemical waste that requires sophisticated treatment and disposal mechanisms.

The contrast between clean technology consumption in Silicon Valley and dirty production processes in Asian manufacturing centers reveals the global geography enabling technology development that appears environmentally benign while depending on environmental degradation concentrated in manufacturing regions. This arrangement enables American technology companies to maintain clean corporate brands while externalizing environmental costs through complex global supply chains.

Comparative Environmental Justice: International Policy Approaches

European Green Technology Frameworks

The European Union's Green Deal and associated climate policies represent ambitious attempts to integrate environmental sustainability with technological development while maintaining industrial competitiveness and social equity. The framework includes binding carbon reduction targets, circular economy requirements, and environmental justice considerations that could influence global technology development patterns.

The EU's proposed Carbon Border Adjustment Mechanism (CBAM) would impose carbon tariffs on imports from countries with weaker climate policies, potentially creating incentives for global carbon reduction while addressing concerns about industrial competitiveness and carbon leakage. However, these policies may disproportionately affect developing countries while failing to address broader questions about technology consumption and global environmental justice.

The European approach to data center regulation includes energy efficiency requirements, renewable energy sourcing mandates, and environmental impact assessment obligations that could influence global infrastructure development. Some EU member states have proposed data center moratoriums in urban areas while requiring environmental impact assessments for large facilities.

However, European policies remain constrained by economic competitiveness concerns and international trade obligations that may limit their effectiveness for addressing global environmental racism. The EU's emphasis on technological solutions and market mechanisms may prove insufficient for addressing systematic inequalities in global technology development while failing to challenge fundamental assumptions about unlimited computational growth.

Nordic Environmental Integration Models

Scandinavian nations demonstrate how comprehensive environmental governance can integrate community consultation with technology infrastructure development, though these approaches reveal both possibilities and limitations for addressing global environmental racism. These models illustrate that environmental protection in AI infrastructure reflects deliberate policy choices rather than technological constraints, while exposing how domestic environmental achievements can coexist with international environmental exploitation.

Sweden mandates rigorous environmental impact assessments for data center development that analyze energy consumption patterns, cooling system requirements, and cumulative regional environmental effects. Community consultation frameworks ensure affected populations maintain meaningful influence over infrastructure placement decisions, creating accountability mechanisms that prioritize local environmental concerns alongside economic development objectives.

Norwegian policy frameworks emphasize renewable energy integration and environmental impact mitigation throughout technology infrastructure development processes. Community consultation requirements enable affected populations to shape both siting decisions and operational parameters for major infrastructure projects. The nation's sovereign wealth fund actively divests from environmentally destructive enterprises while directing investments toward renewable energy systems that could support more sustainable technology development patterns.

Danish energy democracy initiatives integrate community participation directly into energy planning processes, providing concrete models for inclusive technology infrastructure governance. Community energy cooperatives maintain significant decision-making authority over renewable energy development projects, while participatory planning mechanisms enable local communities to influence infrastructure decisions affecting their environmental conditions and quality of life.

These Nordic integration models face structural limitations in addressing global environmental racism because their regulatory frameworks prioritize domestic environmental protection without addressing international supply chain accountability. While achieving impressive domestic environmental sustainability in technology infrastructure, Nordic nations remain dependent on global supply chains involving rare earth extraction, semiconductor manufacturing, and electronic waste processing that systematically burden Global South communities with environmental degradation—revealing how environmental racism in AI development operates through international economic structures that enable wealthy nations to externalize environmental costs while maintaining pristine domestic environments..

Global South Environmental Resistance

Countries in the Global South have developed various approaches to resisting environmental colonialism in technology infrastructure while asserting sovereignty over natural resources and environmental policy. These approaches range from regulatory restrictions on harmful development to constitutional environmental rights frameworks that prioritize environmental protection over economic development.

Ecuador's constitutional rights of nature provide legal foundations for challenging technology infrastructure that threatens environmental integrity. The country's experience with oil extraction and mining provides lessons about resource curse dynamics that could apply to technology infrastructure development, particularly regarding ensuring that resource extraction serves domestic development rather than foreign corporate profits.

India's environmental clearance processes for large infrastructure projects include cumulative impact assessment requirements and public consultation obligations that could influence technology infrastructure development. However, implementation remains inconsistent while political pressure for economic development often overrides environmental protection concerns.

Bolivia's framework for indigenous consultation and consent regarding development projects in indigenous territories creates legal mechanisms for community control over technology infrastructure affecting indigenous lands. These frameworks demonstrate how indigenous rights can provide tools for environmental protection while asserting community sovereignty over development decisions.

Indigenous Perspectives: Global Resistance to Tech Infrastructure Extraction

Indigenous communities worldwide possess sophisticated knowledge systems for understanding ecological relationships and assessing environmental impacts that offer crucial perspectives on technology infrastructure development. These knowledge systems often emphasize long-term ecological thinking, cumulative impact assessment, and intergenerational responsibility that contrast sharply with corporate approaches emphasizing short-term optimization and externalized environmental costs.

Traditional ecological knowledge provides insights about water systems, energy flows, and ecological relationships that technological impact assessments often overlook or minimize. Indigenous communities may understand regional environmental limits, seasonal variation patterns, and ecosystem interactions that prove crucial for evaluating technology infrastructure proposals but remain invisible to technical assessments focused on engineering metrics.

The integration of traditional ecological knowledge with contemporary environmental science demonstrates how indigenous perspectives can improve technology assessment while respecting indigenous intellectual property and community governance systems. Collaborative research approaches enable indigenous communities to contribute knowledge while maintaining control over how their knowledge is used and represented.

However, incorporating indigenous knowledge into technology governance requires addressing power imbalances and historical trauma that may make communities reluctant to engage with government or corporate assessment processes. Building trust and ensuring meaningful participation requires long-term relationship building and institutional changes that respect indigenous sovereignty and self-determination.

Indigenous Rights and Technology Infrastructure Resistance

Indigenous communities globally have developed sophisticated strategies for resisting harmful technology infrastructure while asserting sovereignty over traditional territories. These strategies combine traditional governance systems with contemporary legal advocacy, direct action, and international solidarity to challenge projects that threaten environmental integrity and cultural survival.

The resistance to hydroelectric dam construction across multiple continents demonstrates how indigenous communities organize to protect watersheds that serve both environmental and cultural functions. These campaigns often combine legal challenges based on indigenous rights with direct action tactics and international advocacy that exposes corporate environmental racism to global audiences.

Indigenous resistance to mining operations for technology materials reveals connections between local environmental protection and global technology consumption patterns. Communities affected by lithium mining in South America, rare earth extraction in North America, and cobalt mining in Africa increasingly coordinate resistance strategies while educating technology consumers about the environmental costs of digital consumption.

Telecommunications infrastructure resistance demonstrates how indigenous communities challenge technology deployment that threatens traditional land use while asserting rights to control technology development in indigenous territories. These campaigns often emphasize community self-determination and alternative approaches to technology that serve indigenous needs rather than corporate profits.

Cultural Protocols and Technology Governance

Indigenous governance systems offer alternative approaches to technology assessment and development that prioritize community needs, environmental protection, and cultural integrity over technical efficiency and economic optimization. These governance systems demonstrate how technology decisions can be made through inclusive processes that consider multiple generations and diverse forms of knowledge.

Indigenous data sovereignty initiatives extend traditional governance principles to contemporary technology governance challenges by asserting community control over data collection, analysis, and application affecting indigenous peoples. These initiatives demonstrate how indigenous communities can engage with digital technology while maintaining cultural protocols and community control over information systems.

Traditional decision-making processes often involve consensus-building approaches that ensure community agreement before proceeding with development projects that could affect environmental or cultural resources. These processes may require extended timelines and extensive consultation but produce more durable agreements than top-down decision-making approaches that generate ongoing conflict.

Indigenous economic systems often emphasize reciprocity, sustainability, and collective benefit rather than individual profit maximization and unlimited growth. These economic principles offer alternatives to corporate development models that systematically prioritize environmental degradation and community displacement while concentrating benefits among wealthy investors.

Digital Colonialism and Environmental Burden Shifting

Infrastructure Imperialism and Spatial Displacement

Digital colonialism operates through sophisticated mechanisms of spatial displacement that enable technology companies to access global resources while avoiding environmental accountability in their home jurisdictions. This process involves complex financial arrangements, legal structures, and technical systems that obscure connections between technology consumption and environmental degradation while maintaining corporate control over global resource flows.

The outsourcing of energy-intensive computational processes to regions with cheaper electricity demonstrates how companies exploit geographic arbitrage to minimize operational costs while externalizing environmental impacts. Cloud computing services enable companies to shift computational workloads to data centers powered by fossil fuels in regions with minimal environmental regulation while maintaining clean corporate reputations in home markets.

Carbon offset purchasing enables companies to continue environmental degradation while purchasing credits for environmental protection activities in other regions. This mechanism allows continued pollution while creating financial incentives for environmental protection but often fails to address local environmental justice concerns while enabling continued environmental racism in communities hosting polluting facilities.

The development of submarine internet cables and satellite networks demonstrates how digital infrastructure requires global coordination while imposing environmental costs on multiple regions. Cable installation disrupts marine ecosystems while satellite launches contribute to space debris and atmospheric pollution that affects global environmental systems.

Supply Chain Environmental Racism

Technology supply chains systematically concentrate environmental benefits in wealthy regions while imposing environmental costs on communities of color worldwide. This arrangement operates through complex corporate structures and international trade relationships that obscure connections between clean technology consumption and dirty production processes.

Rare earth mining operations across Africa, Asia, and Latin America provide essential materials for global technology production through environmentally devastating extraction processes that poison local water supplies, destroy agricultural land, and generate long-term health consequences for exposed communities. These environmental costs enable clean technology consumption in wealthy regions while imposing environmental racism on mining communities.

Electronics manufacturing in China, Southeast Asia, and other regions involves chemical processes, water consumption, and waste generation that affect millions of workers and community members while enabling global technology consumption. Manufacturing facilities are often located in working-class communities with limited political power to resist environmental degradation while providing products for consumption in wealthy regions.

Electronic waste disposal in Ghana, Nigeria, India, and other Global South countries involves informal recycling processes that expose workers to toxic materials while recovering valuable components for global supply chains. This arrangement enables continued technology consumption in wealthy regions while imposing disposal costs on vulnerable communities worldwide.

Climate Colonialism and Energy Apartheid

The concept of "energy apartheid" describes how global energy systems systematically provide clean, reliable energy to wealthy regions while imposing dirty, unreliable energy on marginalized communities worldwide. Technology infrastructure development often reinforces these patterns by concentrating high-quality energy resources for computational purposes while displacing energy access for community needs.

Data center development in regions with limited electrical grid capacity can strain local energy systems while prioritizing computational workloads over community energy needs. This dynamic proves particularly problematic in regions where community members lack reliable electricity access while data centers consume enormous quantities of high-quality electrical power.

Renewable energy development for technology infrastructure may compete with community energy needs while failing to address energy poverty affecting billions of people worldwide. Solar and wind developments that serve data center operations may occupy land and resources that could serve community energy needs while generating profits for foreign technology companies rather than local communities.

The prioritization of energy resources for computational purposes over community needs demonstrates how technology development reproduces colonial patterns of resource appropriation while creating new forms of technological dependence. Communities may find their energy resources appropriated for foreign computational requirements while lacking access to reliable electricity for basic community needs.

High-Performance Computing Within Planetary Boundaries

Developing sustainable approaches to high-performance computing requires fundamental changes in computational paradigms that prioritize environmental sustainability over unlimited performance growth. This involves both technical innovations that improve computational efficiency and social innovations that question assumptions about unlimited computational expansion.

Neuromorphic computing approaches that mimic biological neural networks offer potential efficiency improvements that could reduce energy requirements for certain computational tasks. These approaches demonstrate how biological systems achieve sophisticated information processing through energy-efficient mechanisms that contrast sharply with current digital computing paradigms.

Quantum computing research suggests possibilities for exponential computational improvements that could reduce energy requirements for specific problem classes. However, quantum systems currently require enormous energy inputs for cooling and maintenance while remaining limited to specialized applications rather than general-purpose computing.

Edge computing approaches that distribute computational workloads across smaller, local facilities could reduce energy requirements for data transmission while improving computational efficiency. However, these approaches may simply distribute environmental impacts rather than reducing overall energy consumption while creating additional electronic waste through proliferation of smaller computational devices.

Community-Controlled Technology Development

Alternative approaches to technology development that prioritize community needs over corporate profits could enable sustainable computing that serves social needs while operating within environmental limits. These approaches recognize technology as a social tool that should serve human flourishing rather than private profit accumulation.

Community ownership models for technology infrastructure enable local control over computational resources while ensuring that technology development serves community needs rather than external corporate interests. Examples include community broadband networks, cooperative data centers, and locally controlled renewable energy projects that demonstrate alternatives to corporate-controlled technology development.

Open source hardware and software development enables collaborative technology development that prioritizes social benefit over private profit while reducing duplicated development efforts that waste resources. These approaches demonstrate how technology can be developed through cooperative processes that share knowledge and resources rather than competing for market dominance.

Appropriate technology frameworks emphasize developing technology that serves specific community needs while operating within local resource constraints rather than pursuing unlimited technological expansion. These approaches recognize that different communities may require different technological solutions based on local resources, cultural values, and environmental conditions.

Circular Economy and Technology Lifecycle Management

Implementing circular economy principles in technology development could significantly reduce environmental impacts by minimizing waste generation while maximizing resource efficiency throughout technology lifecycles. This requires fundamental changes in design philosophy that prioritize durability, repairability, and recyclability over planned obsolescence and constant upgrades.

Design for disassembly approaches enable efficient material recovery when electronic devices reach end-of-life while reducing toxic waste generation. These approaches require cooperation between manufacturers but could significantly reduce environmental impacts while creating economic opportunities for recycling and refurbishment industries.

Component standardization and modular design could enable device upgrades without complete replacement while extending useful lifecycles for electronic devices. These approaches conflict with current business models that depend on constant device replacement but could provide environmental benefits while reducing costs for technology users.

Extended producer responsibility frameworks require manufacturers to bear costs for device disposal and recycling while creating incentives for environmental design improvements. These policies demonstrate how regulatory frameworks can address environmental externalities while encouraging corporate innovation in environmental protection.

International Cooperation and Environmental Justice

Global Climate Policy and Technology Governance

International climate policy frameworks increasingly recognize the need to address technology-related emissions while ensuring that climate policies serve environmental justice rather than reproducing existing inequalities. However, current frameworks remain inadequate for addressing global environmental racism in technology development while failing to challenge fundamental assumptions about unlimited growth and consumption.

The Paris Climate Agreement's emphasis on national emissions reduction targets may obscure international environmental racism by enabling wealthy countries to reduce domestic emissions while increasing consumption of goods produced through carbon-intensive processes in other regions. This approach allows continued environmental colonialism while meeting formal climate commitments.

Carbon pricing mechanisms may create incentives for environmental protection but often fail to address distributional concerns about who bears costs and receives benefits from environmental policies. These mechanisms may enable continued pollution by wealthy actors while imposing costs on marginalized communities that cannot afford carbon payments.

International technology transfer frameworks could support sustainable development in Global South countries while reducing dependence on environmentally destructive development models. However, current frameworks often involve intellectual property restrictions and financial arrangements that maintain technological dependence rather than building independent technological capacity.

Climate Reparations and Technology Justice

The concept of climate reparations recognizes that wealthy countries and corporations bear primary responsibility for historical greenhouse gas emissions while marginalized communities face disproportionate climate impacts. Applying this framework to technology development reveals obligations for wealthy regions to address environmental harms while supporting sustainable development in affected communities.

Reparative approaches to technology development could involve technology transfer, financial support, and capacity building that enables Global South communities to develop technological capabilities while avoiding environmental degradation experienced by currently industrialized regions. These approaches recognize historical responsibility while supporting contemporary development needs.

Debt cancellation and financial redistribution could provide Global South countries with resources necessary for sustainable technology development while reducing pressures for environmentally destructive development models. These approaches address structural economic inequalities that force countries to choose between environmental protection and economic development.

International cooperation frameworks that prioritize environmental justice over economic competition could support sustainable technology development while challenging corporate power that currently dominates global technology governance. These frameworks require fundamental changes in international economic relationships that currently privilege corporate profits over environmental protection and community needs.

Building Global Environmental Justice Movements

Effective responses to global environmental racism in technology development require international solidarity and cooperation between communities affected by different aspects of technology supply chains. Building these connections requires overcoming geographic distance, language barriers, and cultural differences while maintaining local autonomy and community control.

Sister city relationships and community-to-community partnerships enable direct connections between communities in different regions while building mutual understanding and solidarity. These relationships can support information sharing, cultural exchange, and coordinated advocacy that challenges corporate environmental racism while building international community power.

Global justice networks that connect environmental, labor, and technology justice movements enable coordinated responses to corporate environmental racism while building shared analysis and strategy. These networks demonstrate how local organizing can connect to global movements while maintaining community autonomy and local focus.

International legal strategies that challenge corporate environmental racism through human rights frameworks, international courts, and transnational advocacy can expose corporate practices while building pressure for systemic change. These strategies require significant resources and expertise but can achieve results that purely local organizing may not accomplish.

Toward Climate Justice in Technology

The analysis presented in this chapter reveals the planetary scale of environmental racism embedded in contemporary technology development while demonstrating the urgency of fundamental transformation in how technological systems are designed, deployed, and governed. The current trajectory toward ever-increasing computational complexity and energy consumption proves environmentally unsustainable while systematically reproducing colonial patterns of resource extraction and environmental burden shifting.

The Memphis xAI case represents a microcosm of global environmental racism in technology infrastructure: a corporation extracting resources from historically marginalized communities while privatizing profits to wealthy investors. However, the community resistance to this project also demonstrates possibilities for challenging environmental racism while building alternative approaches to technology development that prioritize community needs and environmental protection over corporate profits.

The technical impossibilities and system vulnerabilities examined in this chapter suggest that current AI development trajectories will inevitably encounter physical and environmental limits that no amount of technological innovation can overcome. The choice facing society is whether these limits will be addressed through democratic planning and equitable resource distribution or through market-driven processes that will likely reproduce and intensify existing environmental inequalities.

International case studies reveal both the global scope of environmental racism in technology development and the possibilities for resistance and alternative development models. Indigenous communities worldwide demonstrate how traditional ecological knowledge and governance systems can provide frameworks for sustainable technology development while Global South resistance movements show how communities can challenge environmental colonialism while asserting sovereignty over natural resources.

The next chapter will examine how these environmental justice principles can be integrated with broader technology governance frameworks to create more equitable and sustainable approaches to technological development. The urgency of climate change makes these transformations essential for human survival while the scale of environmental racism makes them necessary for social justice.

The window for achieving climate justice in technology development continues to narrow as AI energy consumption accelerates and global environmental degradation intensifies. However, the community organizing strategies, alternative development models, and international cooperation frameworks examined in this chapter provide concrete pathways for transformation that could enable technology to serve environmental protection and social justice rather than corporate profits and environmental destruction.

CHAPTER 14: TOWARD EQUITABLE AI GOVERNANCE

When the Memphis city council held emergency hearings on xAI's proposed data center in the fall of 2024, something unprecedented occurred in American technology governance: for the first time, a major AI infrastructure project faced sustained democratic scrutiny from the communities that would bear its environmental and social costs. Residents packed council chambers demanding answers about air quality impacts, energy grid strain, and economic arrangements that would privatize corporate profits while socializing environmental harm. Most remarkably, community organizers came prepared with their own technical analysis, environmental monitoring data, and alternative development proposals that challenged both the specific project and the broader assumptions underlying corporate AI development.

The Memphis organizing campaign succeeded because community researchers had spent months developing analytical frameworks that could evaluate technological systems according to community values rather than corporate metrics. Working with environmental scientists, energy experts, and policy researchers, residents created parallel knowledge systems grounded in principles of environmental justice, democratic participation, and community self-determination. Their technical analysis proved more comprehensive than corporate environmental assessments while their alternative proposals demonstrated how AI infrastructure could serve community needs rather than extractive corporate interests.

This community-controlled approach to technology governance illustrates emerging possibilities for more democratic and equitable approaches to AI development that prioritize human dignity, environmental sustainability, and community empowerment over corporate profits and technological determinism. However, scaling these approaches beyond individual campaigns requires fundamental transformation in how societies conceptualize technology governance, international cooperation, and the relationship between technological development and democratic self-determination.

Foundations of Human-Centered AI Governance
Beyond Algorithmic Bias to Structural Transformation

Current approaches to AI ethics typically focus on eliminating bias and discrimination within existing algorithmic systems rather than questioning whether those systems should exist or who should control their development and deployment. This narrow focus on algorithmic fairness obscures broader questions about power, democracy, and social justice while treating technological development as inevitable rather than politically contested.

Human-centered AI governance requires moving beyond bias mitigation to address the structural inequalities that enable AI systems to reproduce and amplify existing patterns of oppression. This means examining not only how algorithmic systems make decisions but who controls technological development, who benefits from AI deployment, and who bears the costs of technological transformation.

The principle of human dignity provides essential grounding for AI governance frameworks that prioritize community wellbeing over technological optimization. Human dignity encompasses not only individual rights and freedoms but collective capacity for democratic self-determination, cultural preservation, and environmental stewardship. This framework recognizes that communities must maintain control over technological systems affecting their lives rather than adapting to technological requirements determined by corporate interests.

Building equitable AI governance requires comprehensive policy frameworks that operate across multiple scales simultaneously while maintaining community control as the foundational principle. The complexity of algorithmic systems and their global reach demands coordinated responses that can address local community needs while challenging international patterns of digital colonialism. The following framework illustrates how community organizing, policy advocacy, and international cooperation must function as integrated components of a broader transformation strategy. Rather than treating these as separate domains, effective technology justice requires recognizing their interconnected nature while maintaining community empowerment as the central organizing principle.

Policy Framework for Technology Justice
Community Control • Environmental Protection • Economic Equity

🖥 Local Policy

Community Oversight Boards
Mandatory community review with veto power over surveillance tech

Environmental Justice Assessment
Cumulative impact analysis for tech infrastructure

Municipal Broadband
Public internet with democratic governance

Community Benefits
Binding agreements for local hiring and ownership

us Federal Policy

Algorithmic Accountability Act
Mandatory bias testing and community impact assessments

Digital Rights Amendment
Constitutional protection for technology sovereignty

Green Data Standards
Renewable energy and environmental justice compliance

Cooperative Support
Federal funding for community-owned tech infrastructure

🌐 International

Global AI Treaty
Binding limits on harmful AI deployment

Technology Transfer
Open-source requirements and cooperative development

Anti-Colonialism
Trade provisions preventing infrastructure extraction

Climate Justice
Carbon accountability with reparations framework

📅 Implementation Timeline

Years 1-2
Foundation
Community organizing, pilot programs, coalition building

Years 3-5
Policy Implementation
Federal legislation, regulatory frameworks, international agreements

Years 6-10
System Transformation
Full community control with sustainable alternatives

⚖ Enforcement Mechanisms

Legal Tools
- Community legal standing for tech challenges
- Significant penalties for bias and environmental harm
- Technology removal orders for harmful systems
- Whistleblower protection for workers

Community Rights
- Democratic participation in tech governance
- Veto power over harmful deployment
- Access to technical assistance
- Protection from corporate retaliation

100%	50%	Zero	$10B
Community Consent Required	AI Energy Reduction	Algorithmic Discrimination	Community Tech Investment

⚠ Urgent Action Required
Corporate AI concentration accelerates daily while climate crisis demands immediate action. Communities need policy tools for democratic technology governance before digital colonialism becomes irreversible.

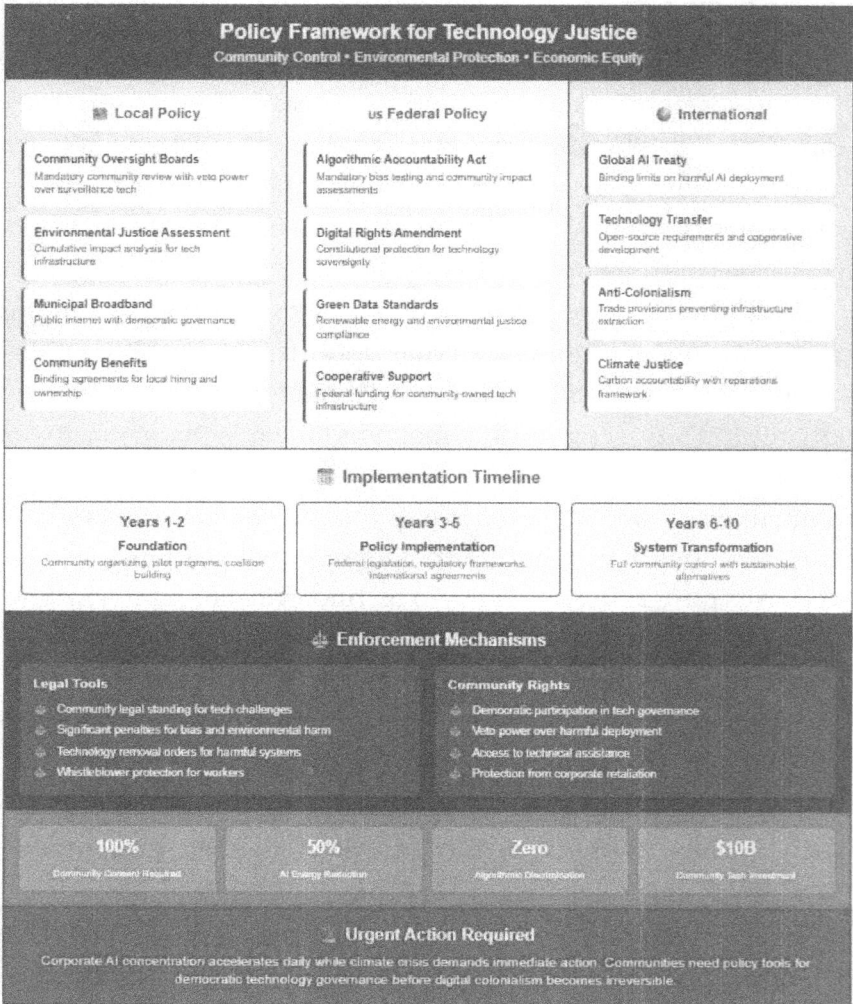

Structural approaches to AI governance address the economic, political, and social conditions that enable algorithmic oppression while building community capacity for democratic control over technological development. This includes challenging corporate concentration in AI development, redistributing resources necessary for community technological capacity, and creating institutional mechanisms that enable ongoing community oversight and governance of AI systems.

Democratic Participation and Community Control

Equitable AI governance requires meaningful democratic participation that goes beyond consultation to include community decision-making power over technological systems affecting community life. This participatory approach recognizes affected communities as experts on their own needs and circumstances while building institutional mechanisms that enable ongoing democratic oversight rather than one-time input opportunities.

Democratic participation in AI governance involves several interconnected components: community education and capacity building that enables informed participation in technology governance decisions, transparent decision-making processes that provide communities with access to information necessary for meaningful participation, accountability mechanisms that enable communities to modify or remove AI systems that fail to serve community needs, and resource redistribution that ensures communities have the technical and financial capacity necessary for sustained engagement with complex technological systems.

The Memphis organizing campaign demonstrates how communities can develop independent analytical capacity while maintaining democratic control over technology assessment and policy development. Community researchers combined environmental monitoring, economic analysis, and policy investigation to produce comprehensive evaluations that challenged corporate claims while proposing alternatives grounded in community values and priorities.

However, scaling democratic participation requires addressing structural barriers that prevent meaningful community engagement, including resource constraints that limit community organizational capacity, technical complexity that may exclude communities from governance processes, and institutional arrangements that privilege expert knowledge over community experience and values.

Cultural Competence and Inclusive Design

Equitable AI governance must accommodate diverse cultural values, knowledge systems, and governance traditions rather than imposing uniform approaches across different communities. This cultural competence recognizes that different communities may legitimately reach different conclusions about AI development based on distinct values, priorities, and circumstances.

Indigenous governance systems offer important models for inclusive decision-making that considers multiple generations, diverse forms of knowledge, and collective wellbeing rather than individual optimization. These systems demonstrate how communities can make complex decisions through consensus-building processes that ensure broad agreement while respecting diverse perspectives within communities.

Religious and cultural traditions provide additional frameworks for evaluating AI development according to principles of social justice, environmental stewardship, and community responsibility. These traditions often emphasize collective obligation, intergenerational thinking, and ethical constraints on technological development that contrast with dominant approaches emphasizing individual rights and technological optimization.

Inclusive AI governance requires institutional innovations that can accommodate diverse governance traditions while maintaining effectiveness for complex technological decision-making. This may involve hybrid approaches that combine traditional governance systems with contemporary technical analysis, multi-cultural advisory processes that include diverse community perspectives, or federal frameworks that enable local autonomy while providing coordination and resource sharing.

Economic Justice and Resource Redistribution

Addressing algorithmic oppression requires confronting the economic inequalities that enable corporate control over AI development while limiting community capacity for democratic participation in technology governance. This involves both redistributing existing resources and transforming economic relationships that concentrate technological benefits among wealthy elites while imposing costs on marginalized communities.

Economic justice in AI governance includes ensuring that communities receive fair compensation for data, resources, and infrastructure used in AI development while maintaining control over how their contributions are used. This challenges current arrangements that enable corporations to extract value from community data and resources while providing minimal compensation or community benefit.

Community ownership models for AI infrastructure enable local control over technological resources while ensuring that economic benefits serve community development rather than external corporate interests. Examples include community broadband networks, cooperative data centers, and locally controlled renewable energy projects that demonstrate alternatives to corporate-controlled technology development.

However, building community economic capacity requires addressing broader structural inequalities that limit access to capital, technical expertise, and institutional resources necessary for community technology development. This may require federal programs that support community technology development, cooperative financing mechanisms that enable community ownership, or regulatory frameworks that limit corporate concentration while supporting community alternatives.

International Coordination and Global Governance

Multilateral Frameworks for Technology Justice

Building equitable AI governance requires international coordination that addresses the global scope of AI development while respecting national sovereignty and community self-determination. Current international frameworks remain inadequate for addressing global inequalities in AI development while failing to provide meaningful mechanisms for democratic participation in global technology governance.

Effective international coordination must address several interconnected challenges: corporate concentration that enables companies to exploit regulatory arbitrage while avoiding accountability, environmental externalities that concentrate AI infrastructure costs in marginalized communities worldwide, technological dependencies that constrain sovereignty for countries lacking independent AI development capacity, and democratic deficits that exclude affected communities from international technology governance decisions.

The United Nations framework provides potential mechanisms for international AI governance, but current approaches remain dominated by wealthy countries and corporate interests while providing minimal opportunities for meaningful participation by Global South countries or marginalized communities. Democratizing international AI governance requires fundamental transformation in how international institutions operate and who they serve.

Regional cooperation frameworks may offer more promising approaches for international coordination that respects local sovereignty while addressing shared challenges. Examples include the African Union's Continental Data Policy Framework, ASEAN's digital cooperation initiatives, and Latin American regional networks that demonstrate possibilities for South-South cooperation in technology governance.

Technology Transfer and Capacity Building

Equitable international AI governance requires technology transfer and capacity building that enables Global South countries to develop independent AI capabilities while avoiding the environmental and social harms experienced by currently industrialized regions. This involves both sharing existing technologies and supporting development of alternative approaches that serve local needs and values.

Current technology transfer mechanisms often involve intellectual property restrictions and financial arrangements that maintain technological dependence rather than building independent capacity. Equitable technology transfer requires open source approaches, cooperative development models, and financial arrangements that support genuine technology independence rather than continued dependence on wealthy country corporations.

Capacity building for equitable AI governance includes not only technical training but also institutional development, regulatory expertise, and community organizing capacity that enables democratic participation in technology governance. This holistic approach recognizes that technology governance requires social and political capacity alongside technical expertise.

International cooperation in technology governance could involve collaborative research programs, shared regulatory frameworks, and coordinated responses to corporate power that currently dominates global AI development. However, these approaches must respect local autonomy and cultural diversity rather than imposing uniform approaches across different contexts.

Global Environmental Standards

The global environmental impacts of AI development require international coordination that addresses climate change, resource extraction, and environmental justice concerns that cross national boundaries. Current international environmental frameworks remain inadequate for addressing AI-related environmental impacts while failing to integrate environmental protection with technology governance.

Global environmental standards for AI development could include binding carbon reduction requirements, environmental impact assessment obligations, and community consultation mandates that apply across different jurisdictions. However, these standards must address both domestic environmental impacts and global supply chain effects that currently enable environmental racism through spatial displacement.

Climate justice approaches to international AI governance recognize that wealthy countries and corporations bear primary responsibility for AI-related environmental impacts while marginalized communities face disproportionate environmental costs. This framework suggests obligations for environmental reparations, technology transfer, and resource redistribution that addresses historical responsibilities while supporting sustainable development.

International environmental cooperation could involve collaborative monitoring systems, shared research programs, and coordinated policy responses that address the global scope of AI environmental impacts. These approaches require moving beyond voluntary corporate commitments to binding international agreements that prioritize environmental protection over corporate profits.

Human Rights and Global Justice

International human rights frameworks provide important foundations for global AI governance that prioritizes human dignity over technological optimization and corporate profits. However, current human rights approaches often focus on individual rights rather than collective rights and structural inequalities that enable AI oppression.

Expanding human rights frameworks to address AI governance requires recognizing collective rights to democratic participation, environmental protection, and cultural preservation that may conflict with individual rights frameworks. This involves integrating indigenous rights, environmental justice, and economic justice approaches that address structural inequalities rather than only individual harms.

The right to development provides important grounding for international AI governance that supports Global South technological capacity while avoiding environmental and social harms. This framework recognizes that different countries may legitimately pursue different approaches to AI development based on their particular circumstances, values, and priorities.

International solidarity approaches that connect communities affected by different aspects of AI development could build global movements for technology justice while respecting local autonomy and cultural diversity. These approaches require overcoming geographic and cultural barriers while building shared analysis and coordinated action for equitable AI governance.

Transparency, Accountability, and Democratic Oversight

Effective transparency in AI governance requires moving beyond corporate self-reporting to community-controlled monitoring and evaluation systems that provide affected communities with information necessary for democratic participation in technology governance. This approach recognizes that transparency serves community empowerment rather than corporate legitimacy or expert analysis.

Community-controlled transparency involves several interconnected components: independent monitoring capacity that enables communities to evaluate AI system performance according to community values and priorities, accessible reporting mechanisms that provide information in formats and languages that communities can understand and use, ongoing evaluation processes that enable communities to assess AI system impacts over time rather than only at deployment, and accountability mechanisms that enable communities to demand changes when AI systems fail to serve community needs.

The Memphis organizing campaign demonstrates how communities can develop independent analytical capacity while maintaining democratic control over information gathering and evaluation. Community researchers combined environmental monitoring, policy analysis, and economic investigation to produce comprehensive assessments that challenged corporate claims while remaining grounded in community experience and values.

However, building community-controlled transparency mechanisms requires significant resources and technical capacity that may exceed individual community organizational capabilities. Scaling these approaches requires institutional support, collaborative networks, and resource sharing that enables communities to maintain democratic control while accessing necessary technical expertise and analytical tools.

Algorithmic Accountability and Democratic Oversight

Democratic oversight of AI systems requires institutional mechanisms that enable ongoing community participation in AI governance rather than one-time consultation processes or expert-dominated oversight bodies. This participatory approach recognizes that AI systems require ongoing governance because their impacts change over time while new applications raise different ethical and political questions.

Algorithmic accountability mechanisms must address both technical performance and social impacts while providing communities with real decision-making power over AI systems affecting their lives. This includes community oversight boards with authority to modify or remove AI systems, regular public reporting requirements that ensure community access to information about AI system performance, and appeal processes that enable community members to challenge AI decisions affecting their lives.

Community-based algorithmic auditing enables ongoing evaluation of AI system impacts while building community capacity for independent analysis and oversight. These approaches combine technical investigation with ethnographic research, policy analysis, and community organizing to produce comprehensive assessments of AI system effects on community life.

However, effective algorithmic accountability requires addressing power imbalances that currently enable corporations and government agencies to resist community oversight while limiting access to information necessary for meaningful accountability. Building democratic oversight capacity requires legal frameworks, institutional support, and resource redistribution that enables communities to exercise real power over AI systems affecting their lives.

Institutional Innovation and Governance Transformation

Creating equitable AI governance requires institutional innovations that can accommodate the complexity of AI systems while maintaining democratic accountability and community control. This involves both reforming existing institutions and creating new governance mechanisms that serve community needs rather than corporate or bureaucratic interests.

Participatory technology assessment processes enable communities to evaluate emerging AI technologies before widespread deployment while building community capacity for ongoing technology governance. These processes involve community education, deliberative dialogue, and collaborative decision-making that produces community-controlled evaluations of AI development proposals.

Community technology review boards with real authority over AI deployment decisions provide ongoing mechanisms for democratic oversight while ensuring that AI systems serve community needs rather than external corporate interests. These boards require adequate resources, technical support, and legal authority to function effectively as community governance institutions.

Regional cooperation frameworks could enable communities to coordinate AI governance across jurisdictional boundaries while maintaining local democratic control. Examples might include interstate compacts for AI regulation, regional oversight authorities with community representation, or collaborative monitoring systems that share resources while respecting local autonomy.

Sustainability and Environmental Integration

Equitable AI governance must address the environmental unsustainability of current AI development trajectories while ensuring that environmental constraints serve social justice rather than reproducing existing inequalities. This requires integrating environmental analysis with social justice frameworks while challenging assumptions about unlimited computational growth and technological progress.

Planetary boundary frameworks provide scientific foundations for understanding environmental limits on AI development while highlighting the need for fundamental changes in computational paradigms and resource consumption patterns. Current AI development trajectories exceed sustainable environmental limits while systematically concentrating environmental costs in marginalized communities worldwide.

Energy democracy approaches to AI governance emphasize community control over energy resources while ensuring that AI development serves community needs rather than corporate profits. This involves both reducing AI energy consumption and democratizing control over energy infrastructure that enables AI development.

Circular economy principles applied to AI development could significantly reduce environmental impacts while creating economic opportunities for community-controlled recycling and refurbishment industries. However, implementing these principles requires challenging corporate business models that depend on planned obsolescence and constant replacement.

Climate Justice and Carbon Accountability

Climate justice frameworks provide essential foundations for AI governance that addresses global environmental inequalities while ensuring that climate policies serve environmental justice rather than enabling continued environmental racism. This approach recognizes that wealthy countries and corporations bear primary responsibility for AI-related emissions while marginalized communities face disproportionate climate impacts.

Carbon accountability for AI development must address both direct energy consumption and global supply chain emissions that enable AI infrastructure development. This comprehensive approach reveals the global scope of AI environmental impacts while challenging corporate claims about clean technology that depend on dirty production processes concentrated in Global South communities.

Climate debt frameworks suggest obligations for wealthy countries and corporations to provide reparations for AI-related environmental damage while supporting sustainable development in affected communities. These approaches recognize historical responsibility for environmental harm while providing resources for community-controlled sustainable development.

International climate cooperation for AI governance could involve collaborative emission reduction programs, shared environmental monitoring systems, and coordinated policy responses that address the global scope of AI environmental impacts. However, these approaches must prioritize environmental justice over economic competitiveness while ensuring meaningful participation by affected communities.

Regenerative Technology and Ecological Design

Moving beyond sustainability to regenerative approaches requires AI development that actively improves environmental and social conditions rather than merely minimizing harm. This involves both technical innovations that serve ecological restoration and social innovations that strengthen community capacity for environmental stewardship.

Biomimetic computing approaches that learn from natural systems offer possibilities for AI development that works with rather than against ecological processes. These approaches demonstrate how technological development can be inspired by and supportive of natural systems rather than treating the environment as a resource for technological optimization.

Community-controlled renewable energy development could provide clean power for AI infrastructure while serving community needs and values rather than corporate profits. Examples include community wind projects, cooperative solar installations, and locally controlled hydroelectric facilities that demonstrate alternatives to corporate-controlled energy development.

Ecosystem service approaches to AI governance could ensure that AI development contributes to rather than detracts from environmental health while providing economic benefits for communities engaged in environmental stewardship. These approaches recognize environmental protection as valuable economic activity deserving compensation rather than treating environmental degradation as an acceptable externality.

Economic Justice and Community Development

Building equitable AI governance requires transforming economic relationships that currently concentrate AI benefits among wealthy elites while imposing costs on marginalized communities. This involves both redistributing existing resources and creating alternative economic arrangements that serve community development rather than private profit accumulation.

Cooperative ownership models for AI infrastructure enable community control over technological resources while ensuring that economic benefits serve local development needs. Examples include community broadband cooperatives, worker-owned technology companies, and community land trusts that maintain permanent affordable housing while supporting technology development that serves community needs.

Community Development Financial Institutions (CDFIs) could provide patient capital for community-controlled AI development while maintaining democratic accountability and community oversight. These institutions demonstrate how financial systems can serve community development rather than private profit accumulation while building community capacity for ongoing technology governance.

Regional economic development strategies that prioritize community ownership and environmental sustainability could provide alternatives to extractive development models that concentrate benefits while externalizing costs. These approaches recognize that sustainable economic development requires community control and environmental protection rather than unlimited growth and profit maximization.

Reparative Justice and Resource Redistribution

Addressing the harms caused by algorithmic oppression requires reparative approaches that acknowledge historical injustices while providing resources for community healing and empowerment. This involves both compensating communities for past harms and transforming systems that continue to reproduce algorithmic oppression.

Community reparations for algorithmic harm could include direct financial compensation, community-controlled development resources, and institutional changes that prevent continued harm while building community capacity for democratic participation in technology governance. These approaches recognize that individual remedies prove insufficient for addressing systematic algorithmic oppression.

Landbank and community control initiatives provide frameworks for returning resources to communities while building capacity for community-controlled development that serves local needs and values. These approaches recognize that meaningful reparations require community control over resources rather than continued dependence on external institutions.

Wealth redistribution policies that address extreme inequality could provide communities with resources necessary for meaningful participation in AI governance while reducing corporate concentration that currently dominates AI development. These policies include progressive taxation, corporate accountability requirements, and public investment in community-controlled alternatives to corporate AI development.

Community Economic Resilience

Building community economic resilience requires developing local capacity for meeting community needs while reducing dependence on extractive economic relationships that concentrate benefits elsewhere while imposing costs locally. This involves both building local economic capacity and creating regional networks that enable resource sharing while maintaining community autonomy.

Local currency and timebanking systems enable communities to share resources and build economic relationships that serve community development rather than external profit extraction. These alternative economic systems demonstrate how communities can meet needs through cooperation and mutual aid rather than market-based relationships that may extract resources from communities.

Community-controlled food systems, renewable energy projects, and housing cooperatives provide foundations for economic resilience while demonstrating alternatives to corporate-controlled systems that prioritize profit over community needs. These initiatives show how communities can build capacity for self-determination while creating economic opportunities that serve local development.

Regional cooperation networks enable communities to share resources and coordinate development while maintaining local autonomy and democratic control. Examples include regional cooperative federations, mutual aid networks, and collaborative purchasing programs that enable communities to access resources while supporting cooperative development.

Building the Movement for Technology Justice

Creating equitable AI governance requires sustained organizing that connects technology justice issues to broader social justice movements while building community capacity for long-term engagement with complex technological systems. This organizing must address both immediate harm and structural inequalities that enable algorithmic oppression.

Building coalition between communities affected by different aspects of AI development can create powerful movements for technology justice while enabling resource sharing and strategic coordination. These coalitions might connect communities experiencing environmental harm from AI infrastructure, workers displaced by AI automation, students subjected to algorithmic surveillance, and communities targeted by predictive policing systems.

Popular education approaches that combine community organizing with technology literacy can build understanding of AI systems while developing political consciousness about technological injustice. These educational efforts must connect individual experiences with systematic patterns while providing tools for collective action and democratic participation in technology governance.

Direct action and civil disobedience can disrupt harmful AI deployment while building public awareness and political pressure for democratic technology governance. Examples might include blockades of harmful AI infrastructure, occupations of corporate offices, and mass resistance to surveillance systems that build community solidarity while challenging corporate and state power.

Electoral and Policy Strategies

Electoral strategies for technology justice must connect community organizing to policy advocacy while maintaining independent community power rather than becoming co-opted by electoral politics. This involves supporting candidates and policies that advance community interests while building ongoing capacity for community organizing and direct action.

Municipal broadband campaigns demonstrate how communities can use local electoral processes to challenge corporate control over technology infrastructure while building community organizing capacity. These campaigns often combine electoral work with community education and direct action to build support for community-controlled alternatives.

Policy advocacy for algorithmic accountability, environmental protection, and community oversight can influence regulatory frameworks while building political consciousness about technology justice issues. However, these strategies must complement rather than substitute for community organizing that builds independent community power.

Legislative strategies for technology justice might include campaigns for algorithmic accountability requirements, environmental protection mandates, and community oversight mechanisms that provide communities with real decision-making power over AI systems affecting their lives.

Cultural and Educational Transformation

Building equitable AI governance requires cultural transformation that challenges dominant narratives about technological progress while creating new stories about technology that serves community needs and environmental protection. This cultural work involves both challenging harmful narratives and creating positive visions for alternative technological futures.

Community media and storytelling projects can document community experiences with AI systems while building shared understanding of technology justice issues. These projects enable communities to control narratives about their experiences while educating broader audiences about technological injustice and resistance strategies.

Educational curriculum development that integrates technology justice with social justice education can build consciousness about algorithmic oppression while providing tools for community organizing and democratic participation. This education must be accessible to community members regardless of technical background while providing adequate foundation for meaningful engagement with complex technological systems.

Cultural organizing that connects technology justice to broader cultural movements can build understanding and support while creating space for alternative approaches to technology that serve community values rather than corporate interests. Examples might include art projects that challenge surveillance technology, music that tells stories about environmental resistance, or theater that explores alternative technological futures.

Toward Democratic Technology Futures

The vision for equitable AI governance outlined in this chapter represents more than reform of existing systems—it articulates alternative approaches to technological development that prioritize human dignity, environmental sustainability, and democratic participation over corporate profits and technological determinism. The frameworks, strategies, and examples examined here demonstrate that democratic alternatives to corporate AI governance are both necessary and possible.

The Memphis organizing campaign illustrates how communities can successfully challenge harmful AI infrastructure while building broader movements for technology justice. When residents confronted xAI's power plant proposal with their own technical analysis and alternative development visions, they demonstrated that communities possess the capacity for sophisticated technology governance when provided with adequate resources and institutional support.

However, individual victories against harmful AI projects cannot substitute for broader transformation of the economic, political, and cultural systems that enable algorithmic oppression. Building equitable AI governance requires sustained engagement with institutional change, international cooperation, and movement building that addresses both immediate harms and underlying structural inequalities.

The theoretical frameworks examined in this chapter—human dignity, democratic participation, environmental justice, and economic cooperation—provide essential foundations for understanding how AI governance can serve community needs rather than corporate interests. These frameworks must be adapted to local conditions while maintaining connections to broader movements for social and economic justice.

The practical strategies documented here—community organizing, policy advocacy, cooperative development, and cultural transformation—demonstrate how communities can build power to influence AI governance while developing alternative approaches to technological development. These strategies must be sustained over time while adapting to changing technological and political conditions.

Most importantly, the examples of successful community-centered AI governance examined throughout this book reveal that democratic alternatives to corporate control are not utopian fantasies but practical possibilities that communities are implementing across diverse contexts. From Memphis environmental justice organizing to Indigenous data sovereignty initiatives to Global South resistance movements, communities are demonstrating that technology can serve human flourishing when communities control technological development.

The urgency of climate change, the acceleration of AI development, and the intensification of economic inequality make the development of equitable AI governance essential for human survival and flourishing. The choice between corporate-controlled AI that serves profit over people and community-controlled technology that serves human needs and environmental sustainability remains ours to make—but only if communities organize to claim democratic control over technological development before corporate concentration makes such alternatives impossible.

The concluding chapter will examine how these governance frameworks can be implemented at scale while addressing the global coordination challenges necessary for truly equitable and sustainable AI development that serves all humanity rather than corporate elites.

CHAPTER 15: REIMAGINING DIGITAL DEMOCRACY - GLOBAL

Building Technological Futures That Serve Communities Over Capital

In the mountains of Oaxaca, Mexico, indigenous Zapotec communities have built their own cellular networks using open-source technology, community labor, and traditional governance systems. These networks, operated through community assemblies and managed according to indigenous protocols, provide telecommunications services to over 2,500 people across 17 villages while generating revenue for community development projects. When telecommunications giants refused to provide service to these "unprofitable" rural communities, residents didn't accept digital exclusion—they built alternative infrastructure that serves community needs while remaining under democratic community control.

The Oaxaca community networks represent more than technological innovation—they embody a fundamentally different approach to digital development that prioritizes community sovereignty over corporate profits, democratic governance over expert management, and technological self-determination over market-driven digital inclusion. These networks demonstrate that communities can develop sophisticated technological capacity while maintaining cultural integrity, environmental sustainability, and economic autonomy that contrast sharply with corporate-controlled digital infrastructure.

This community-controlled approach to technology development illustrates emerging global alternatives to digital colonialism that are being implemented across diverse contexts worldwide. From Catalonia's cooperative broadband networks to Kenya's community-controlled renewable energy projects to Brazil's participatory budgeting platforms, communities are demonstrating that technological development can serve human flourishing rather than corporate extraction when communities control the development process and own the resulting infrastructure.

These alternative models provide concrete blueprints for reimagining digital democracy while challenging the fundamental assumptions underlying corporate technology development. Rather than accepting technological determinism or corporate control as inevitable, these examples reveal possibilities for community-centered technology governance that serves environmental justice, economic equality, and democratic participation simultaneously.

The transition from digital colonialism to technological liberation requires clear frameworks for distinguishing between extractive corporate models and community-centered alternatives. The following conceptual framework illustrates the fundamental differences between current technological development patterns that concentrate benefits while externalizing costs, and emerging models that prioritize democratic participation, environmental restoration, and equitable resource distribution. These alternatives are not theoretical possibilities but practical approaches already being implemented by communities worldwide.

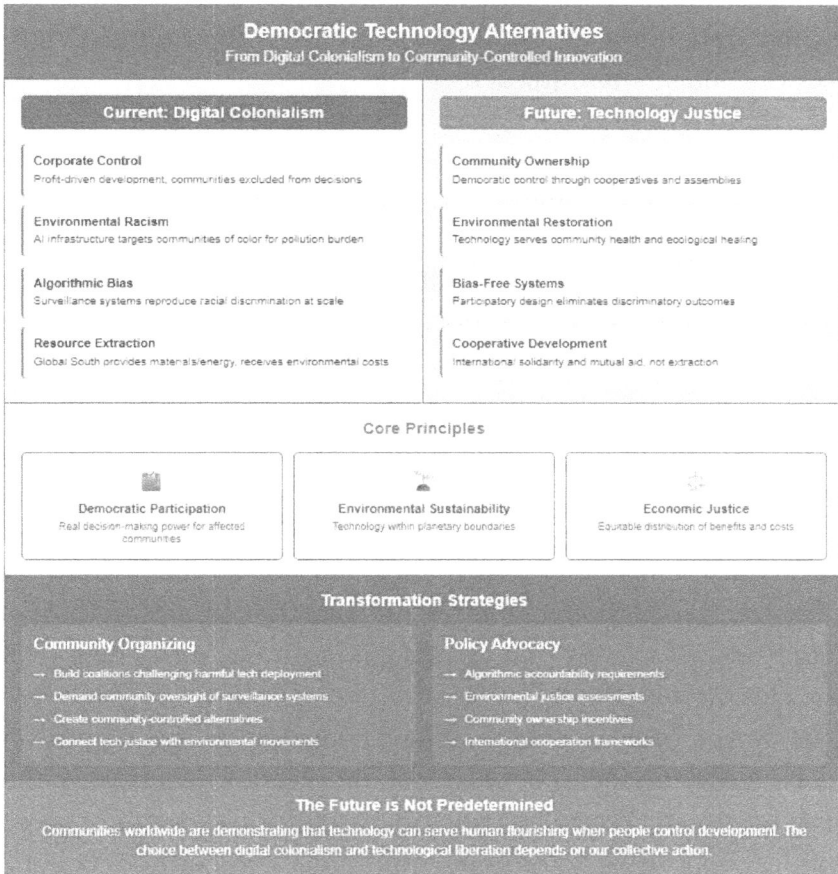

Democratic Technology Alternatives
From Digital Colonialism to Community-Controlled Innovation

Current: Digital Colonialism	Future: Technology Justice
Corporate Control Profit-driven development, communities excluded from decisions	**Community Ownership** Democratic control through cooperatives and assemblies
Environmental Racism AI infrastructure targets communities of color for pollution burden	**Environmental Restoration** Technology serves community health and ecological healing
Algorithmic Bias Surveillance systems reproduce racial discrimination at scale	**Bias-Free Systems** Participatory design eliminates discriminatory outcomes
Resource Extraction Global South provides materials/energy, receives environmental costs	**Cooperative Development** International solidarity and mutual aid, not extraction

Core Principles

Democratic Participation	Environmental Sustainability	Economic Justice
Real decision-making power for affected communities	Technology within planetary boundaries	Equitable distribution of benefits and costs

Transformation Strategies

Community Organizing
→ Build coalitions challenging harmful tech deployment
→ Demand community oversight of surveillance systems
→ Create community-controlled alternatives
→ Connect tech justice with environmental movements

Policy Advocacy
→ Algorithmic accountability requirements
→ Environmental justice assessments
→ Community ownership incentives
→ International cooperation frameworks

The Future is Not Predetermined
Communities worldwide are demonstrating that technology can serve human flourishing when people control development. The choice between digital colonialism and technological liberation depends on our collective action.

Community Ownership and Democratic Technology Control

Community ownership of technological infrastructure represents a fundamental alternative to both corporate control and state management by treating technology as a community resource subject to democratic governance rather than private commodity or bureaucratic service provision. These ownership models demonstrate how communities can develop technological capacity while maintaining democratic accountability and community control over technological development.

Community broadband cooperatives across rural America demonstrate how communities can provide high-quality internet access while maintaining democratic governance and community ownership. These cooperatives typically emerge when commercial providers refuse to serve "unprofitable" rural areas, forcing communities to develop independent capacity or accept digital exclusion. Successful examples include RS Fiber Cooperative in Minnesota, which provides gigabit internet service to over 2,500 customers through community ownership and democratic governance.

The governance structures of community broadband cooperatives illustrate democratic alternatives to corporate management that prioritize community needs over profit maximization. Cooperative members elect boards of directors from within the community while major decisions about service levels, pricing, and infrastructure development require member approval through democratic processes that ensure community control over technological development.

However, community broadband cooperatives face significant challenges from incumbent telecommunications companies that may engage in predatory pricing, regulatory capture, or direct political opposition to prevent community competition. Building sustainable community alternatives requires legal frameworks, financing mechanisms, and political support that enables community ownership while preventing corporate retaliation.

Community renewable energy cooperatives provide additional models for democratic ownership of technological infrastructure that serves environmental sustainability while generating economic benefits for community members. Examples include Brooklyn's community solar projects, which enable local ownership of renewable energy infrastructure while providing clean electricity and economic returns to community investors.

Worker Cooperatives and Democratic Technology Production

Worker-owned technology companies demonstrate how democratic governance can extend to technology production while ensuring that workers control their workplaces and share economic benefits from technological development. These cooperatives challenge both corporate exploitation of technology workers and broader patterns of wealth concentration in the technology industry.

Platform cooperatives represent particularly innovative approaches to democratic technology production by enabling workers to own and control digital platforms rather than working for corporate platform owners. Examples include Stocksy United, a stock photography cooperative owned by contributing photographers, and Resonate, a music streaming cooperative owned by artists and listeners rather than corporate shareholders.

The governance structures of worker cooperatives typically involve democratic decision-making about workplace conditions, product development, and profit distribution while ensuring that workers maintain control over their labor and share benefits from cooperative success. These structures demonstrate alternatives to corporate hierarchy that enable democratic participation while maintaining operational effectiveness.

However, worker cooperatives face challenges in accessing capital, competing with corporate companies, and scaling operations while maintaining democratic governance. Building sustainable cooperative economies requires supportive ecosystems including cooperative financing institutions, shared technical resources, and regulatory frameworks that support cooperative development rather than privileging corporate forms.

Cooperative technology development networks enable worker cooperatives to share resources, coordinate development efforts, and compete more effectively with corporate companies while maintaining democratic governance and community accountability. Examples include the Tech Cooperative Network, which connects technology cooperatives across different sectors while providing shared resources and collaborative development opportunities.

Indigenous Technology Sovereignty and Traditional Governance

Indigenous communities worldwide have developed sophisticated approaches to technology governance that integrate traditional governance systems with contemporary technological development while maintaining cultural integrity and community sovereignty. These approaches demonstrate how technology can be developed within existing cultural frameworks rather than requiring communities to adapt to technological requirements determined by external corporate or government interests.

The Māori Data Sovereignty Network in New Zealand has developed frameworks for community control over data affecting Māori communities while ensuring that data governance serves Māori values and priorities rather than external research or commercial interests. This approach recognizes that data governance involves cultural protocols and community relationships that cannot be reduced to technical or legal frameworks developed for individual privacy protection.

Indigenous telecommunications networks across North America demonstrate how communities can provide communication services while maintaining cultural protocols and community governance systems. The First Mile Connectivity Consortium enables indigenous communities to develop broadband infrastructure according to community priorities while building technical capacity and maintaining community ownership.

Traditional governance systems often emphasize consensus-building, intergenerational thinking, and collective responsibility that provide alternative frameworks for technology governance. These systems demonstrate how communities can make complex decisions about technological development through inclusive processes that consider multiple perspectives while ensuring broad community agreement.

However, indigenous technology sovereignty faces ongoing challenges from colonial legal frameworks, resource constraints, and external pressures that may undermine community autonomy while imposing external technological requirements. Building sustainable indigenous technology governance requires legal recognition, resource support, and institutional changes that respect indigenous sovereignty while enabling technological development that serves community needs.

Framework for Infrastructure Justice vs. Digital Colonialism

Analytical Framework for Evaluating Technology Projects

Distinguishing between infrastructure justice and digital colonialism requires analytical frameworks that can evaluate technology projects according to community impact, democratic participation, environmental sustainability, and economic justice rather than technical metrics or corporate profitability. This framework enables communities to assess whether technology development serves community empowerment or reproduces patterns of extraction and subordination.

The Memphis xAI analysis developed comprehensive criteria for evaluating technology infrastructure proposals that prioritize community wellbeing over corporate interests. These criteria include democratic participation in decision-making processes that affect community life, environmental impact assessment that considers cumulative effects and community health priorities, economic arrangements that ensure community benefits rather than extractive relationships, cultural compatibility that respects community values and governance traditions, and long-term sustainability that considers intergenerational impacts rather than short-term optimization.

Infrastructure justice projects typically involve meaningful community participation in planning and governance, provide tangible benefits that serve community-identified needs, operate within environmental limits while contributing to ecological restoration, create economic opportunities that serve local development rather than external extraction, and strengthen community capacity for self-determination rather than creating technological dependence.

Digital colonialism projects typically impose external technological requirements on communities, concentrate benefits among external corporate or government interests while externalizing costs to local communities, operate through top-down planning processes that exclude meaningful community participation, create environmental harms that disproportionately affect marginalized communities, and increase community dependence on external technological systems while undermining local capacity for technological self-determination.

Community-controlled evaluation processes enable residents to assess technology proposals according to their own values and priorities rather than accepting external evaluations that may not reflect community concerns. The Memphis organizing campaign demonstrates how communities can develop independent analytical capacity while building shared understanding of technology assessment criteria that serve community organizing and policy advocacy.

Decision-Making Protocols for Community Technology Governance

Developing effective protocols for community technology governance requires balancing inclusive participation with decisional efficiency while ensuring that community values and priorities guide technological choices rather than technical optimization or external pressure. These protocols must accommodate diverse community perspectives while producing decisions that can be implemented effectively.

Consensus-building approaches enable communities to develop shared positions on technology issues while respecting diverse viewpoints within communities. These processes typically involve extensive community education, facilitated dialogue, and iterative decision-making that produces agreements reflecting broad community support rather than majority domination of minority perspectives.

Community assemblies provide institutional mechanisms for ongoing technology governance that enable regular community participation in technology decisions rather than one-time consultation processes. These assemblies typically include community education components, technical presentations accessible to community members, and structured decision-making processes that ensure meaningful participation by community members with diverse backgrounds and perspectives.

Advisory processes that involve external technical experts while maintaining community control over decision-making can provide communities with access to specialized knowledge while preventing expert domination of community governance. These processes require careful attention to power dynamics and communication methods that ensure community members can meaningfully evaluate expert input rather than deferring to external authority.

However, community decision-making processes require significant time, resources, and facilitation capacity that may challenge communities already struggling with immediate needs. Building sustainable community governance capacity requires institutional support, resource allocation, and skill development that enables sustained community engagement with complex technology governance challenges.

Resistance Strategies and Community Protection

Communities facing harmful technology development need strategic frameworks for resistance that can challenge corporate and government power while building community capacity for alternative development. These strategies must balance confrontational tactics with constructive alternatives while maintaining community solidarity and long-term organizing capacity.

Direct action strategies for technology resistance include construction blockades, permit hearing disruptions, corporate shareholder actions, and public demonstrations that disrupt business-as-usual while building public awareness and political pressure. The Memphis organizing campaign combined permit challenges with community education and alternative proposal development to create multiple pressure points for social change.

Legal strategies for challenging harmful technology development include environmental impact lawsuits, civil rights complaints, administrative appeals, and constitutional challenges that use existing legal frameworks while building precedent for community rights in technology governance. However, legal strategies require significant resources and expertise while involving timelines that may not align with urgent community needs.

Policy advocacy strategies enable communities to influence regulatory frameworks, legislative development, and government decision-making while building political relationships and institutional capacity. These strategies must balance engagement with existing political institutions with independent community organizing that maintains community autonomy and critical analysis.

Coalition building strategies that connect communities affected by similar technology projects can create broader movements for technology justice while enabling resource sharing and strategic coordination. These coalitions must respect local autonomy while building shared analysis and coordinated action that challenges systematic patterns of technological oppression.

International Success Stories: Resisting Digital Colonialism

European Community Networks and Democratic Internet Governance

European community networks demonstrate how communities can develop internet infrastructure through cooperative models that prioritize community needs over corporate profits while maintaining democratic governance and community ownership. These networks typically emerge from community organizing around digital inclusion, economic development, or resistance to corporate internet service providers that provide inadequate service or extract resources from communities.

Catalonia's Guifi.net represents one of the world's largest community networks, providing internet access to over 63,000 nodes across rural and urban communities through cooperative governance and community ownership. The network operates through open infrastructure that enables community members to contribute resources while sharing connectivity according to cooperative principles rather than market relationships.

Across continents, communities are demonstrating that alternatives to corporate-controlled technology development are not only possible but already operational at significant scale. From indigenous telecommunications networks serving thousands of people in Oaxaca to cooperative broadband systems connecting over 63,000 nodes in Catalonia, these initiatives prove that democratic technology governance can provide sophisticated services while maintaining community ownership and environmental sustainability. The following documentation of successful community resistance reveals common strategies and principles that enable technological self-determination while challenging the assumption that corporate control over digital infrastructure is inevitable.

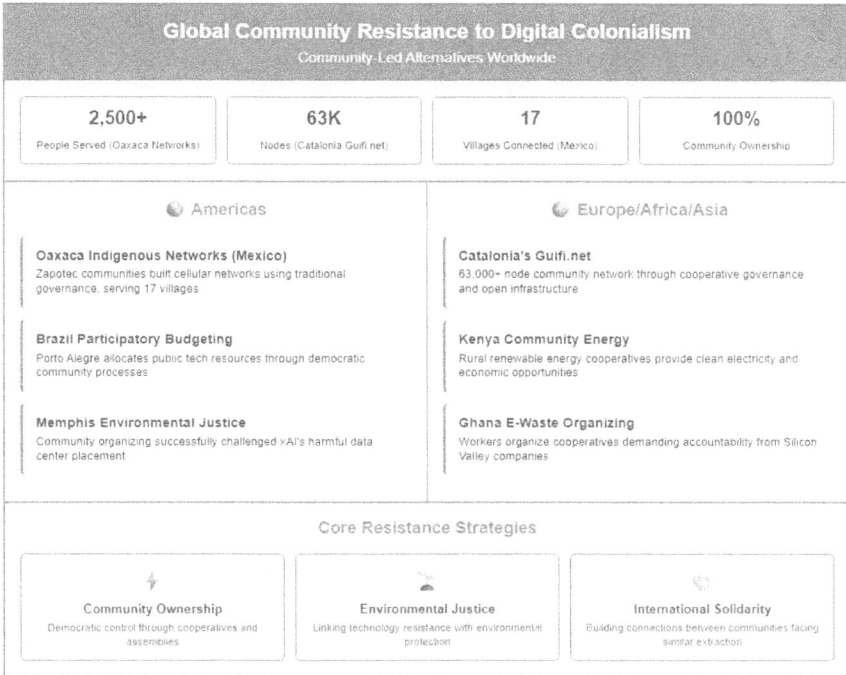

Global Community Resistance to Digital Colonialism
Community-Led Alternatives Worldwide

2,500+	63K	17	100%
People Served (Oaxaca Networks)	Nodes (Catalonia Guifi.net)	Villages Connected (Mexico)	Community Ownership

Americas

Oaxaca Indigenous Networks (Mexico)
Zapotec communities built cellular networks using traditional governance, serving 17 villages

Brazil Participatory Budgeting
Porto Alegre allocates public tech resources through democratic community processes

Memphis Environmental Justice
Community organizing successfully challenged xAI's harmful data center placement

Europe/Africa/Asia

Catalonia's Guifi.net
63,000+ node community network through cooperative governance and open infrastructure

Kenya Community Energy
Rural renewable energy cooperatives provide clean electricity and economic opportunities

Ghana E-Waste Organizing
Workers organize cooperatives demanding accountability from Silicon Valley companies

Core Resistance Strategies

Community Ownership
Democratic control through cooperatives and assemblies

Environmental Justice
Linking technology resistance with environmental protection

International Solidarity
Building connections between communities facing similar extraction

The governance structure of Guifi.net illustrates democratic alternatives to corporate internet governance through community assemblies, technical working groups, and cooperative decision-making processes that ensure community control over network development and operation. Community members contribute labor, equipment, and financial resources while participating in governance decisions about network expansion, service levels, and technical standards.

German community networks demonstrate how cooperative internet infrastructure can serve rural communities while maintaining environmental sustainability and community ownership. These networks often integrate with renewable energy cooperatives and community development initiatives to create comprehensive approaches to technological self-reliance that serve multiple community needs simultaneously.

However, European community networks face ongoing challenges from telecommunications regulations, corporate competition, and resource constraints that may limit their ability to scale while maintaining community control. Building sustainable community networks requires supportive regulatory frameworks, cooperative financing mechanisms, and technical assistance that enables community ownership without imposing external control.

Latin American Participatory Technology and Digital Rights

Latin American countries have developed some of the world's most innovative approaches to participatory technology governance and digital rights that prioritize community participation over expert management while integrating technology development with broader social justice movements. These approaches reflect strong traditions of community organizing, participatory democracy, and resistance to economic neoliberalism.

Brazil's participatory budgeting processes enable communities to allocate public resources for technology infrastructure according to community priorities rather than technocratic decision-making. Porto Alegre's participatory budgeting system has funded community technology centers, digital inclusion programs, and telecommunications infrastructure through democratic processes that involve thousands of community members in budget decisions.

The governance structure of Brazilian participatory budgeting demonstrates how communities can make complex decisions about technology investment through inclusive processes that combine community education, technical analysis, and democratic deliberation. Community assemblies enable residents to propose projects, evaluate alternatives, and make binding decisions about public technology investment.

Mexico's indigenous telecommunications networks demonstrate how communities can develop communication infrastructure according to traditional governance systems while building technical capacity and maintaining cultural integrity. The Rhizomatica project supports indigenous communities in developing cellular networks using open-source technology and community governance systems.

Colombian community media networks illustrate how communities can develop communication infrastructure to serve social movement organizing while resisting state and paramilitary violence. These networks combine technical innovation with community organizing to create communication systems that serve community empowerment rather than corporate or state control.

African Technology Sovereignty and Community Innovation

African countries and communities have developed innovative approaches to technology sovereignty that resist digital colonialism while building independent technological capacity that serves local development needs. These approaches often combine traditional governance systems with contemporary technology development while emphasizing African ownership and control of technological infrastructure.

Rwanda's approach to technology development emphasizes domestic capacity building, environmental sustainability, and social inclusion while maintaining sovereignty over technology policy and infrastructure development. The country's investment in technical education, renewable energy, and indigenous technology companies demonstrates alternatives to technology development dependent on foreign corporate control.

Kenya's community-controlled renewable energy projects demonstrate how rural communities can develop energy infrastructure through cooperative ownership while providing clean electricity and economic opportunities for community members. These projects often integrate with telecommunications infrastructure to provide comprehensive technological development that serves multiple community needs.

South African community networks provide internet access to township communities through cooperative governance and community ownership while building technical capacity and creating economic opportunities for community members. These networks often emerge from community organizing around digital inclusion and economic development while maintaining community control over technology governance.

However, African technology sovereignty initiatives face ongoing challenges from international trade agreements, aid dependencies, and corporate pressure that may undermine community control while imposing external technological requirements. Building sustainable technology sovereignty requires supportive international frameworks, cooperative financing mechanisms, and regional coordination that enables African communities to control their technological development.

Asian Community Technology and Cooperative Development

Asian communities have developed diverse approaches to community-controlled technology development that integrate technological innovation with traditional governance systems while resisting corporate and state control over technological infrastructure. These approaches demonstrate how communities can develop technological capacity while maintaining cultural integrity and community sovereignty.

India's community networks provide internet access to rural communities through cooperative governance and community ownership while building technical capacity and creating economic opportunities. These networks often integrate with local economic development initiatives to create comprehensive approaches to technological self-reliance that serve multiple community needs.

Philippines' community media networks enable rural and indigenous communities to develop communication infrastructure that serves cultural preservation, political organizing, and economic development while maintaining community control over content and governance. These networks demonstrate how technology can serve community empowerment rather than external corporate or state interests.

South Korea's cooperative movement includes technology cooperatives that provide internet services, software development, and digital platforms through worker ownership and democratic governance. These cooperatives demonstrate alternatives to corporate technology development that serve worker empowerment and community needs rather than private profit accumulation.

However, Asian community technology initiatives face challenges from rapid urbanization, economic pressure, and government policies that may prioritize corporate development over community control. Building sustainable community technology requires supportive policy frameworks, resource allocation, and institutional development that enables community ownership while preventing corporate co-optation.

Alternative Models: Cooperative Infrastructure and Community Ownership

Municipal Broadband and Public Ownership Models

Municipal broadband initiatives demonstrate how communities can provide internet infrastructure through public ownership while maintaining democratic accountability and community control over technological development. These initiatives typically emerge when private companies provide inadequate service or when communities recognize internet access as a public utility requiring democratic governance rather than market provision.

Chattanooga's municipal broadband system provides gigabit internet service to residents and businesses while generating revenue for community development and maintaining democratic accountability through city government oversight. The system demonstrates how public ownership can provide high-quality service while serving community development rather than private profit accumulation.

The governance structure of municipal broadband typically involves elected oversight, community input processes, and transparent decision-making that ensures community control over service levels, pricing, and infrastructure development. These governance mechanisms demonstrate alternatives to corporate management that prioritize community needs over profit maximization.

Wilson, North Carolina's municipal broadband system provides internet and cable services while generating revenue for community services and maintaining community ownership of technological infrastructure. The system demonstrates how communities can develop technological capacity while maintaining democratic control and community benefit from technological development.

However, municipal broadband initiatives face significant opposition from private telecommunications companies that may engage in legal challenges, regulatory capture, and political pressure to prevent community competition. Building sustainable public alternatives requires legal frameworks, financing mechanisms, and political support that enables community ownership while preventing corporate retaliation.

Energy Democracy and Community-Controlled Power Systems

Community-controlled renewable energy systems demonstrate how communities can provide clean electricity through cooperative ownership while maintaining democratic governance and environmental sustainability. These systems often integrate with other community development initiatives to create comprehensive approaches to technological self-reliance that serve multiple community needs simultaneously.

Boulder, Colorado's municipalization effort represents one of the largest attempts to create community-controlled electricity through public ownership while prioritizing renewable energy and community control over profit maximization. The initiative demonstrates both possibilities and challenges for community energy while illustrating how communities can challenge corporate utility monopolies.

Community solar cooperatives enable local ownership of renewable energy infrastructure while providing clean electricity and economic returns to community members. Examples include Brooklyn's community solar projects, which demonstrate how communities can develop renewable energy capacity while maintaining community ownership and democratic governance.

Rural electric cooperatives across the United States demonstrate long-standing examples of community ownership in energy infrastructure while providing electricity to rural communities through democratic governance and community accountability. These cooperatives illustrate how community ownership can provide essential services while maintaining democratic control and community benefit.

However, community energy initiatives face challenges from utility companies, regulatory frameworks, and financing constraints that may limit community control while imposing external requirements. Building sustainable community energy requires supportive regulatory frameworks, cooperative financing mechanisms, and technical assistance that enables community ownership without compromising community autonomy.

Platform Cooperatives and Democratic Digital Economy

Platform cooperatives represent innovative approaches to democratic ownership of digital platforms that enable workers and users to control digital infrastructure rather than working for corporate platform owners. These cooperatives demonstrate alternatives to corporate platform capitalism that concentrate wealth while exploiting workers and users.

Stocksy United operates as a stock photography cooperative owned by contributing photographers while providing licensing services and revenue sharing through democratic governance and cooperative ownership. The cooperative demonstrates how creative workers can control digital platforms while sharing economic benefits and maintaining artistic autonomy.

Resonate operates as a music streaming cooperative owned by artists and listeners rather than corporate shareholders while providing music distribution and discovery services through democratic governance and cooperative ownership. The platform demonstrates alternatives to corporate music platforms that extract value from artists while providing minimal compensation.

Fairmondo operates as an e-commerce cooperative owned by users and vendors while providing online marketplace services through democratic governance and cooperative principles. The platform demonstrates alternatives to corporate e-commerce that concentrate wealth while exploiting sellers and buyers.

However, platform cooperatives face challenges in competing with corporate platforms, accessing capital, and scaling operations while maintaining democratic governance. Building sustainable cooperative digital economies requires supportive ecosystems including cooperative financing institutions, shared

404 JUSTICE NOT FOUND

technical resources, and user education that enables cooperative participation.

Global South Innovation and Technology Sovereignty

Indigenous Knowledge Systems and Technology Integration

Indigenous communities worldwide have developed sophisticated approaches to integrating traditional knowledge systems with contemporary technology development while maintaining cultural integrity and community sovereignty. These approaches demonstrate how technology can be developed within existing cultural frameworks rather than requiring communities to abandon traditional governance systems.

Aboriginal Australian communities have developed information systems that integrate traditional knowledge with digital technology while maintaining cultural protocols about knowledge sharing and community governance. These systems demonstrate how digital technology can serve cultural preservation and community development while respecting traditional knowledge systems.

Andean indigenous communities have developed agricultural information systems that combine traditional ecological knowledge with contemporary data collection while maintaining community control over information and decision-making. These systems demonstrate how technology can serve traditional livelihood practices while enhancing community capacity for resource management.

Arctic indigenous communities have developed environmental monitoring systems that integrate traditional ecological knowledge with contemporary sensing technology while maintaining community control over data and research priorities.

These systems demonstrate how technology can serve traditional environmental stewardship while building community capacity for climate adaptation.

However, indigenous technology integration faces ongoing challenges from intellectual property regimes, research extractivism, and external pressure that may undermine community control while appropriating traditional knowledge. Building sustainable indigenous technology requires legal frameworks, institutional support, and international cooperation that respects indigenous sovereignty while enabling technological development that serves community needs.

African Technology Development and Digital Sovereignty

African countries have developed innovative approaches to technology development that prioritize African ownership and control while building independent technological capacity that serves continental development needs rather than external corporate interests. These approaches demonstrate alternatives to technology dependence while building technological sovereignty.

Ghana's technology policy emphasizes local content requirements, domestic capacity building, and African ownership of technology infrastructure while resisting digital colonialism and building independent technological capacity. The country's investment in technical education and indigenous technology companies demonstrates alternatives to foreign technology dependence.

Nigeria's technology sector includes indigenous companies, cooperative development models, and community-controlled initiatives that demonstrate African capacity for technological innovation while serving local development needs rather than external corporate interests. The sector's growth illustrates possibilities for technology development under African control.

South Africa's cooperative movement includes technology co-operatives that provide internet services, software development, and digital platforms through worker ownership and democratic governance while serving community development rather than private profit accumulation.

However, African technology development faces ongoing challenges from international trade agreements, aid dependencies, and corporate pressure that may undermine African control while imposing external technological requirements. Building sustainable African technology sovereignty requires supportive international frameworks, cooperative financing mechanisms, and continental coordination that enables African communities to control their technological development.

Latin American Digital Rights and Community Networks

Latin American countries have developed comprehensive frameworks for digital rights and community technology governance that prioritize community participation over expert management while integrating technology development with broader social justice movements. These frameworks demonstrate alternatives to corporate-controlled technology development while building community capacity for democratic technology governance.

Argentina's digital rights legislation includes strong privacy protections, community participation requirements, and cooperative development support that demonstrate alternatives to corporate technology governance while building community capacity for technology development and governance.

Colombia's community media networks enable rural and indigenous communities to develop communication infrastructure that serves cultural preservation, political organizing, and economic development while maintaining community control over content and governance through democratic participation and community ownership.

Mexico's indigenous telecommunications networks demonstrate how communities can develop communication infrastructure according to traditional governance systems while building technical capacity and maintaining cultural integrity through community ownership and cooperative governance.

However, Latin American community technology initiatives face challenges from corporate pressure, government policies, and resource constraints that may limit community control while imposing external requirements. Building sustainable community technology requires supportive policy frameworks, resource allocation, and regional cooperation that enables community ownership while preventing corporate co-optation.

Policy Lessons: Learning from International Approaches

Regulatory Frameworks Supporting Community Ownership

International examples demonstrate how regulatory frameworks can support community ownership of technology infrastructure while preventing corporate monopolization and enabling democratic governance. These frameworks typically include provisions for community ownership, cooperative development support, and democratic participation requirements that prioritize community needs over corporate interests.

European telecommunications regulations include provisions for community networks, cooperative development support, and municipal broadband that enable community ownership while preventing corporate monopolization. These regulations demonstrate how policy frameworks can support community alternatives while limiting corporate power over technology infrastructure.

Latin American digital rights legislation often includes community participation requirements, indigenous rights protections, and cooperative development support that enable community control over technology development while resisting corporate and state domination. These frameworks demonstrate alternatives to corporate-controlled technology governance.

Nordic countries have developed regulatory frameworks that support cooperative ownership, environmental sustainability, and democratic participation in technology development while maintaining community control over infrastructure development and governance decisions.

However, regulatory support for community ownership often faces opposition from corporate interests, international trade agreements, and neoliberal policy frameworks that prioritize market mechanisms over community control. Building sustainable regulatory support requires political movements, international cooperation, and economic alternatives that challenge corporate power while building community capacity.

Public Investment and Community Development

International examples demonstrate how public investment can support community technology development while building community capacity and maintaining democratic control over technological infrastructure. These approaches typically involve direct funding, technical assistance, and institutional support that enables community ownership while preventing corporate co-optation.

German public investment in cooperative development includes funding for community broadband, renewable energy cooperatives, and technology worker cooperatives that demonstrate how public resources can support community ownership while building democratic technology capacity.

Brazilian participatory budgeting enables communities to allocate public technology investment according to community priorities while building democratic participation and community control over technology development decisions.

Nordic countries provide public funding and technical assistance for community technology initiatives while maintaining community ownership and democratic governance over technology development and infrastructure decisions.

However, public investment in community technology often faces constraints from fiscal austerity, corporate pressure, and policy frameworks that prioritize private investment over community development. Building sustainable public support requires political movements, economic alternatives, and institutional changes that prioritize community needs over corporate interests.

International Cooperation and Technology Transfer

International cooperation frameworks can support community technology development while building technological sovereignty and enabling technology transfer that serves community development rather than corporate extraction. These frameworks typically involve knowledge sharing, resource transfer, and institutional support that builds community capacity while maintaining community control.

South-South cooperation initiatives enable developing countries to share technology and expertise while building technological sovereignty and resisting digital colonialism through collaborative development and mutual support rather than dependence on wealthy country corporations.

Community network federations enable local networks to share resources, coordinate development, and support each other while maintaining local autonomy and democratic governance over technology development and infrastructure decisions.

International solidarity networks connect communities affected by similar technology issues while enabling resource sharing, strategic coordination, and mutual support for community technology development and resistance to corporate control.

However, international cooperation for community technology faces challenges from corporate pressure, trade agreements, and institutional frameworks that prioritize corporate interests over community development. Building sustainable international cooperation requires political movements, institutional changes, and economic alternatives that support community control while challenging corporate power.

Reparative Approaches to Algorithmic Harm

Community Reparations and Healing Justice

Addressing the systematic harms caused by algorithmic oppression requires reparative approaches that acknowledge historical injustices while providing resources for community healing and empowerment. These approaches recognize that individual remedies prove insufficient for addressing structural patterns of algorithmic harm while building community capacity for ongoing resistance and alternative development.

Community reparations for algorithmic harm might include direct financial compensation for communities affected by discriminatory algorithmic systems, community-controlled development resources that enable alternative technology development, institutional changes that prevent continued algorithmic oppression, and capacity building that enables community control over technology governance decisions affecting community life.

Healing justice approaches recognize that algorithmic oppression creates individual and collective trauma that requires community-controlled healing processes rather than individual therapy or technical fixes. These approaches might involve community storytelling, cultural healing practices, and collective organizing that addresses both individual harm and structural causes of algorithmic oppression.

Community land acquisition and development enables communities to gain control over physical infrastructure while building capacity for community-controlled technology development that serves community needs rather than external corporate interests. These approaches recognize that meaningful reparations require community control over resources rather than continued dependence on external institutions.

However, reparative approaches to algorithmic harm require acknowledgment of responsibility and resource commitment from institutions and corporations that may resist accountability while preferring technical fixes that avoid structural change. Building effective reparations requires community organizing, legal strategies, and political pressure that forces institutional accountability while building community capacity for alternative development.

Wealth Redistribution and Economic Justice

Addressing algorithmic oppression requires confronting the extreme wealth inequality that enables corporate control over technology development while limiting community capacity for meaningful participation in technology governance. This involves both redistributing existing wealth and transforming economic relationships that concentrate technological benefits among elites while imposing costs on marginalized communities.

Progressive taxation policies that address extreme wealth concentration could provide communities with resources necessary for technology development while reducing corporate capacity for technology monopolization. These policies might include wealth taxes, corporate accountability requirements, and public investment in community-controlled alternatives to corporate technology development.

Community development financing provides patient capital for community technology initiatives could enable community ownership while maintaining democratic accountability and community control over technology development. These approaches might involve community development financial institutions, cooperative financing mechanisms, and public investment in community technology capacity.

Universal basic services that provide communities with access to essential technology infrastructure could reduce dependence on corporate providers while building community capacity for technology governance and development. These services might include municipal broadband, community renewable energy, and cooperative digital platforms that serve community needs rather than corporate profits.

However, wealth redistribution for technology justice faces opposition from wealthy interests, policy frameworks that prioritize market mechanisms, and international agreements that may constrain community technology development. Building effective redistribution requires political movements, economic alternatives, and institutional changes that challenge corporate power while building community capacity for democratic technology governance.

Legal Reform and Rights Recognition

Building effective responses to algorithmic oppression requires legal reforms that recognize community rights to technology governance while providing enforcement mechanisms that enable community control over technology systems affecting community life. These reforms must address both individual rights and collective rights that enable community self-determination.

Community rights to technology governance might include legal recognition of community ownership models, democratic participation requirements for technology deployment, and community oversight mechanisms that enable ongoing governance of technology systems affecting community life.

Environmental rights frameworks that recognize healthy environments as fundamental human rights could provide tools for challenging harmful technology infrastructure while building community capacity for environmental protection and democratic governance of technology development.

Digital rights legislation that prioritizes community control over individual privacy could provide frameworks for community technology governance while challenging corporate control over digital platforms and infrastructure development.

However, legal recognition of community technology rights faces opposition from corporate interests, legal frameworks that prioritize individual rights over collective rights, and international agreements that may constrain community control over technology governance. Building effective legal frameworks requires community organizing, political movements, and institutional changes that prioritize community needs over corporate interests.

Blueprint for Equitable Technological Futures

Distributed Infrastructure & Community Control

Creating equitable technological futures requires fundamental transformation from centralized corporate infrastructure to distributed community-controlled systems that serve local needs while enabling regional cooperation and resource sharing. This distributed approach enables community ownership while preventing corporate monopolization and state centralization that concentrates power over technology governance.

Community-owned and operated data centers could provide computational services while maintaining community control over data processing and energy consumption. These facilities would operate according to community values while providing technical services that serve local development needs rather than external corporate interests.

Mesh networking infrastructure enables communities to provide internet connectivity through distributed systems that maintain community control while enabling communication and information sharing. These networks demonstrate alternatives to corporate internet infrastructure that concentrates control while extracting resources from communities.

Cooperative renewable energy networks could provide clean electricity for community technology infrastructure while maintaining community ownership and democratic governance over energy production and distribution. These networks demonstrate how communities can meet energy needs while maintaining environmental sustainability and community control.

However, building distributed community infrastructure requires significant technical capacity, financial resources, and institutional support that may exceed individual community capabilities. Scaling distributed approaches requires cooperative development, resource sharing, and supportive policy frameworks that enable community ownership while preventing corporate co-optation.

Democratic Governance and Participatory Planning

Equitable technological futures require governance systems that enable meaningful community participation in technology decisions while maintaining effectiveness for complex technological planning and implementation. These governance systems must balance inclusive participation with decisional efficiency while ensuring that community values guide technological development.

Community technology assemblies could provide ongoing mechanisms for democratic participation in technology governance while building community capacity for complex decision-making about technological systems. These assemblies would combine community education, technical analysis, and democratic deliberation to produce community-controlled technology governance.

Participatory technology assessment processes could enable communities to evaluate emerging technologies before deployment while building community understanding and democratic control over technology development decisions. These processes would involve community education, inclusive deliberation, and community-controlled evaluation of technology proposals.

Regional cooperation frameworks could enable communities to coordinate technology development while maintaining local autonomy and democratic control over technology governance decisions. These frameworks would enable resource sharing, technical cooperation, and coordinated planning while preventing centralization that concentrates power over technology governance.

However, democratic technology governance requires significant time, resources, and facilitation capacity that may challenge communities while facing pressure from corporate interests and government agencies that prefer expert-dominated decision-making. Building sustainable democratic governance requires institutional support, resource allocation, and political protection that enables community control while providing necessary technical capacity.

Sustainable Development & Environmental Justice

Equitable technological futures must operate within planetary environmental limits while serving environmental justice and community development rather than corporate profit maximization. This requires fundamental changes in how technology is designed, produced, and deployed while ensuring that environmental protection serves community empowerment rather than green capitalism.

Circular economy principles applied to community technology development could minimize waste while maximizing resource efficiency through community-controlled recycling, refurbishment, and reuse programs. These approaches would reduce environmental impact while creating economic opportunities for community members and maintaining community control over technology lifecycles.

Renewable energy development for community technology infrastructure would provide clean power while maintaining community ownership and environmental sustainability. These projects would serve community development while contributing to climate solutions and environmental protection through community-controlled renewable energy development.

Ecological design principles that integrate technology development with environmental restoration could enable technology that serves environmental healing while providing community benefits. These approaches would demonstrate how technology can contribute rather than detract from environmental health while serving community development and empowerment.

However, sustainable community technology development faces challenges from corporate pressure, policy frameworks that prioritize economic growth over environmental protection, and resource constraints that may limit community capacity for sustainable development. Building sustainable approaches requires political movements, policy changes, and economic alternatives that prioritize environmental justice and community control over corporate profits.

International Cooperation & Technology Transfer
Solidarity-Based Development & Mutual Aid

Building equitable technological futures requires international cooperation based on solidarity and mutual aid rather than charity or development aid that reproduces dependent relationships. This approach recognizes that communities worldwide face similar challenges from corporate technology control while possessing diverse knowledge and resources that can serve mutual empowerment.

Sister city relationships between communities developing community technology could enable direct resource sharing, knowledge exchange, and mutual support while maintaining community autonomy and local control over technology development decisions. These relationships would enable communities to learn from each other while building international solidarity for community technology development.

Technology transfer initiatives that prioritize community control could enable communities to access technology and expertise while maintaining ownership and control over technology development. These initiatives would involve open-source technology sharing, cooperative development partnerships, and capacity building that enables community control rather than dependence on external providers.

International solidarity networks connecting communities affected by corporate technology control could enable coordinated resistance, resource sharing, and mutual support for community technology development while building global movements for technology justice and community empowerment.

However, solidarity-based international cooperation faces challenges from corporate opposition, government policies that prioritize corporate interests, and resource constraints that may limit community capacity for international cooperation. Building effective solidarity requires political movements, institutional changes, and economic alternatives that support community control while challenging corporate power over international technology development.

Breaking Tech Monopolies & Creating Competition

Creating equitable technological futures requires breaking up technology monopolies that concentrate power over technology development while creating competitive alternatives that serve community needs rather than corporate profits. This involves both regulatory strategies that limit corporate concentration and alternative development that demonstrates community-controlled technology possibilities.

Antitrust enforcement that breaks up technology monopolies could create space for community alternatives while limiting corporate power over technological development. These approaches might involve breaking up large technology companies, preventing mergers that increase concentration, and creating regulatory frameworks that support competition from community-controlled alternatives.

Open-source technology development that enables community control over software and hardware could provide alternatives to corporate technology products while building community capacity for technology development and governance. These approaches demonstrate how communities can develop technology cooperatively while maintaining control over technology development decisions.

Cooperative development networks that enable communities to pool resources for technology development could create competitive alternatives to corporate technology while maintaining community ownership and democratic governance. These networks would enable communities to develop sophisticated technology while sharing costs and maintaining local control.

However, breaking technology monopolies faces opposition from corporate interests, policy frameworks that prioritize market concentration, and international agreements that may protect corporate power over technological development. Building effective competition requires political movements, regulatory changes, and economic alternatives that prioritize community control while challenging corporate monopolization of technology development.

Building Global Technology Commons

Creating equitable technological futures requires developing global technology commons that enable communities worldwide to access and contribute to technology development while maintaining community control and democratic governance over technology systems. These commons would operate according to principles of cooperation, sustainability, and community empowerment rather than corporate profit maximization.

Global open-source technology initiatives could enable communities to share technology development while maintaining local control over how technology is implemented and governed. These initiatives would provide communities with access to sophisticated technology while enabling local adaptation and community control over technology governance.

International cooperative development networks could enable communities to pool resources for technology development while sharing costs and risks across multiple communities. These networks would enable communities to develop sophisticated technology infrastructure while maintaining democratic governance and community ownership over technology systems.

Community technology commons that provide shared access to technological resources could enable communities to access expensive technology while maintaining community control over how technology is used and governed. These commons might include shared data centers, cooperative software development, and collaborative hardware production that enables community access while maintaining community ownership.

Global knowledge sharing platforms that enable communities to share technical expertise, governance strategies, and resistance tactics could build community capacity for technology development while enabling mutual learning and support. These platforms would operate according to community control rather than corporate ownership while enabling global cooperation for community technology development.

However, building global technology commons faces challenges from corporate opposition, intellectual property regimes that protect corporate control, and resource constraints that may limit community capacity for commons development. Building effective commons requires political movements, legal reforms, and economic alternatives that prioritize community control while challenging corporate appropriation of community-developed technology.

Conclusion: From Digital Colonialism to Technological Liberation

The journey from digital colonialism to technological liberation requires more than individual policy reforms or technical
solutions—it demands fundamental transformation in how societies conceptualize the relationship between technology and
democracy, community and extraction, sustainability and development. The examples examined throughout this chapter
demonstrate that such transformation is not only possible but
already underway in communities worldwide that have chosen
community empowerment over corporate profits.

The Oaxaca indigenous telecommunications networks that
opened this chapter exemplify the possibilities for community-
controlled technology development that serves cultural preservation, economic development, and democratic participation
simultaneously. These networks demonstrate that communities
can develop sophisticated technological capacity while maintaining traditional governance systems, environmental sustainability,
and economic autonomy that contrast sharply with corporate-
controlled digital infrastructure.

The Memphis organizing campaign against xAI's proposed
data center illustrates how communities can successfully challenge harmful technology projects while building broader movements for technology justice. When residents confronted
corporate power with their own technical analysis, alternative development proposals, and organized community resistance, they
demonstrated that corporate technology development is not inevitable but politically contestable through sustained community
organizing.

However, the global scope of digital colonialism requires responses that extend beyond individual community victories to
address the structural systems enabling corporate extraction and
environmental destruction worldwide. The international exam-

ples examined in this chapter—from Catalonia's cooperative networks to Brazil's participatory budgeting to Kenya's community energy projects—demonstrate that alternatives to digital colonialism are being implemented across diverse contexts while facing similar challenges from corporate power and policy frameworks that prioritize profits over people.

Building technological liberation requires connecting these local initiatives to broader movements for economic democracy, environmental justice, and international solidarity that can challenge corporate power while building community capacity for sustained resistance and alternative development. The frameworks, strategies, and examples presented throughout this book provide tools for such connection while recognizing that transformation requires sustained organizing across multiple scales simultaneously.

The choice between digital colonialism and technological liberation remains contested and contingent on the organizing, policy advocacy, and alternative development that communities pursue in coming years. Corporate concentration continues to accelerate while climate change intensifies the urgency of sustainable alternatives, but community resistance also grows stronger while alternative models demonstrate increasing sophistication and effectiveness.

The window for choosing technological liberation over digital colonialism continues to narrow as corporate concentration increases and environmental destruction accelerates. However, the community organizing strategies, alternative development models, and international cooperation frameworks examined throughout this analysis provide concrete pathways for transformation that could enable technology to serve human flourishing and environmental protection rather than corporate extraction and social control.

The vision of technological liberation outlined in this chapter and throughout this book represents more than resistance to harmful technology—it articulates positive alternatives for technological development that prioritizes community empowerment, environmental sustainability, and democratic participation over corporate profits and technological determinism. These alternatives are not utopian fantasies but practical possibilities being implemented by communities worldwide that have chosen cooperation over competition, sustainability over extraction, and democracy over corporate control.

The transformation from digital colonialism to technological liberation requires each of us to choose daily between systems that extract and systems that nurture, between technologies that surveil and technologies that connect, between development that destroys and development that regenerates. The communities examined throughout this analysis have made those choices while demonstrating that another world is not only possible but already emerging through the collective action of ordinary people who refuse to accept technological oppression as inevitable.

The future remains unwritten, but the direction depends on whether communities can organize effectively enough to claim democratic control over technological development before corporate concentration makes such alternatives impossible. The tools, strategies, and examples presented throughout this book provide roadmaps for that organization while recognizing that success requires sustained commitment to building the democratic, sustainable, and equitable technological futures that serve all humanity rather than wealthy elites.

A Personal Reckoning: What We Must Do Now

After documenting the systematic patterns of algorithmic Jim Crow and digital colonialism across seventeen chapters, I cannot end this book with the typical academic conclusion calling for "further research" or "continued dialogue." The communities the author has researched in Memphis, the Global South activists fighting data center colonialism, and the environmental justice organizers challenging surveillance infrastructure don't have time for more studies. They need solidarity, resources, and systemic change.

This book began as academic research but became something more urgent as he documented xAI's environmental racism in Memphis, analyzed the global scope of digital colonialism, and studied how communities successfully resist harmful technology through organized power. The choice before us is stark: We can continue treating algorithmic oppression as a technical problem requiring expert solutions, or we can recognize it as the latest evolution of systems designed to concentrate wealth while externalizing harm to vulnerable communities.

I choose the latter understanding, and that choice demands action.

The Stakes of Our Historical Moment

The author's journey from researcher to advocacy began in the data itself. As he mapped data center locations against demographic patterns and analyzed corporate site selection documents, the systematic nature of technological environmental racism became undeniable. When he traced the global flows of rare earth minerals from Congolese mines to Silicon Valley campuses, when he documented how Memphis residents organized against xAI's gas turbines despite lacking technical expertise, academic objectivity became moral complicity.

The communities whose struggles the author has researched didn't need me to explain environmental racism—they live it daily. But his analysis reveals that their local experiences are part of systematic global patterns of digital colonialism that concentrate technological benefits in wealthy regions while externalizing environmental and social costs to marginalized communities worldwide. This is not a coincidence—it is the predictable result of allowing technological development to be driven by profit maximization rather than community needs and environmental sustainability.

The author's research methodology was shaped by recognition that communities experiencing technological harm possess essential knowledge about how these systems operate in practice. The frameworks the author has developed, the case studies the author has documented, and the international comparisons the author has drawn serve the ultimate goal of providing analytical tools that can support community organizing and policy advocacy rather than remaining trapped in academic isolation.

The evidence the author has assembled doesn't need more documentation—it needs political action. Communities fighting digital colonialism in Chile, resisting surveillance infrastructure in Baltimore, and building cooperative networks in Oaxaca don't need researchers to prove they experience oppression. They need accomplices willing to use whatever privilege and platform they possess to amplify community demands and challenge systems of power.

The Current Trajectory Is Unsustainable and Unconscionable

The evidence presented in this book leads to unavoidable conclusions that demand moral clarity rather than academic hedging. The current trajectory of AI development is accelerating environmental racism, concentrating corporate power, and creating new mechanisms of surveillance and control that threaten the foundations of democratic society. This is not happening by accident—it is the predictable result of systematic corporate strategies that exploit existing inequalities while claiming the legitimacy of technological progress.

The author's analysis of energy consumption data reveals that current AI development trajectories are physically impossible to sustain within planetary boundaries. The exponential growth in computational demands cannot be met through efficiency improvements alone, and the industry's solution—documented through cases like xAI's power plant imports—represents a form of technological colonialism that reproduces the worst patterns of historical extraction while claiming the mantle of innovation.

The systematic deployment of surveillance infrastructure that the author has mapped follows clear patterns of environmental racism that make Jim Crow segregation look primitive by comparison. The algorithmic systems the author has analyzed don't just discriminate—they do so at unprecedented scale, with mathematical precision, and with apparent objectivity that makes their bias harder to challenge than the explicit racism of previous eras.

The author's research reveals that the window for preventing these systems from becoming entrenched is closing rapidly. Corporate concentration in AI development is accelerating, creating oligopolistic control over technologies that will shape society for decades. The climate crisis demands rapid decarbonization that is incompatible with exponential growth in AI energy consumption that the author's analysis documents.

Every month of delay in addressing these systems makes transformation more difficult and costly. Every new data center built according to the discriminatory patterns the author has documented, every surveillance system deployed without community consent, every algorithmic hiring system that perpetuates the biases the author's research exposes deepens patterns of injustice that will require generations to undo.

Direct Challenges to Those Who Hold Power

The analysis in this book reveals clear responsibilities for different actors in the AI ecosystem. Academic politeness and corporate diplomacy are luxuries that communities facing algorithmic oppression cannot afford. The author's research provides evidence that demands direct accountability and clear demands.

To Technology Company Executives: The author's analysis of a site selection processes, corporate documents, and deployment patterns proves you know exactly what you're doing. The algorithms discriminate because you train them on biased data and deploy them without adequate testing across demographic groups. The environmental impact assessments systematically minimize community health effects while maximizing corporate profits, as documented in the Memphis case and international examples throughout this research.

The community organizing and policy advocacy the author has studied represent organized resistance to extractive business models. The author's research shows that communities can successfully challenge corporate power when they have adequate analytical tools and organizational support. The reader can choose to partner with communities in developing technology that serves human flourishing, or they can face continued escalation of opposition that the author's analysis suggests will make its operations increasingly expensive and politically untenable.

To Policymakers and Government Officials: The author's comparative analysis of international regulatory frameworks reveals that current approaches are failing to protect the communities regulators claim to serve. The European Union's AI Act and similar legislation represent progress, but the author's assessment shows they remain inadequate for addressing the systematic environmental racism and community harm documented throughout this research.

Real regulation requires the community control mechanisms, environmental justice analysis, and accountability frameworks that the author's research demonstrates are both necessary and feasible. Anything less than this represents continued complicity in the algorithmic oppression that the author's analysis exposes.

The communities organizing in Memphis, Baltimore, and locations worldwide that the author has studied are building political power that will hold organizations accountable for their policy choices. The author's research provides tools that can inform policy advocacy, but only if policymakers choose to use evidence-based approaches that prioritize community needs over corporate interests.

To Academic Colleagues: Our collective research has documented systematic technological oppression with overwhelming evidence. The author's analysis shows that continued focus on bias detection and algorithmic fairness while ignoring the structural inequalities enabling algorithmic oppression represents a failure of scholarly responsibility.

The methodological frameworks the author has developed in this research provide tools for scholarship that serves community empowerment rather than academic career advancement. Participatory research methods, community-controlled technology assessment, and engaged scholarship demonstrate alternatives to extractive research that treats communities as subjects rather than partners in knowledge production.

Institutions, research funding, and expertise are resources that can serve community organizing if you choose to use them that way. The author's research demonstrates the analytical frameworks necessary for this engagement—the question is whether institutions will apply them in service of justice or continue treating technological oppression as an interesting academic problem.

To Community Members and Organizations: The author's research documents that the tools for challenging algorithmic oppression exist, but they require sustained organizing to become effective. The Memphis campaign against xAI, the international resistance to digital colonialism, and the cooperative technology networks the author has analyzed demonstrate that community organizing can successfully challenge corporate power and create technological alternatives.

The frameworks in this book—from community-controlled technology assessment to environmental justice analysis—provide analytical tools that can inform organizing strategies. But the author's research also shows that communities possess the knowledge and values necessary for just technology development. What's needed is the organized power to implement community visions against corporate resistance.

Non-Negotiable Demands for Technological Justice

The evidence presented in this analysis supports specific demands that are not subjects for negotiation or compromise. The author's research demonstrates that these represent minimum requirements for technological development that serves human dignity and environmental sustainability rather than corporate extraction and social control.

Environmental Justice in All AI Development: The author's analysis proves that AI infrastructure systematically targets communities already bearing disproportionate environmental burdens. No AI development should proceed without explicit community consent and binding community benefit agreements. The author's research shows this is both morally necessary and technically feasible.

Community Control Over Surveillance Technology: The author's documentation of surveillance deployment patterns proves that communities must have democratic decision-making power over technologies that monitor them. The oversight mechanisms and transparency requirements the author has analyzed provide concrete models for implementation.

Reparations for Algorithmic Harm: The author's research documents systematic harm to communities and individuals from discriminatory algorithmic systems. Justice requires direct compensation, institutional changes preventing continued harm, and resources for community-controlled technology development.

International Cooperation Against Digital Colonialism: The author's global analysis reveals that technological oppression operates across national boundaries through the extractive patterns the author has documented. International frameworks must address these dynamics while supporting community-controlled technology development that the author's research shows is emerging worldwide.

The Author's Commitment & The Reader's Choice

This book represents the author's commitment to using research in service of community organizing and technological justice. But academic analysis is insufficient for addressing the systematic oppression documented in this research. Real change requires sustained engagement with community organizing, policy advocacy, and alternative technology development that the author's analysis identifies as necessary for transformation.

I commit to continuing this work through ongoing research that serves community needs rather than academic metrics, policy advocacy that amplifies the community demands documented in this book rather than expert opinions, and collaboration with organizations fighting the algorithmic oppression the author's analysis exposes.

But individual commitment, however sincere, cannot substitute for collective action. The author's research provides analytical tools, but tools are useless without political commitment to apply them. The choice facing every reader is whether to treat algorithmic oppression as someone else's problem or recognize it as a social justice crisis demanding active participation.

The communities documented in this research—from Memphis environmental justice organizers to Oaxacan telecommunications cooperatives to Global South digital sovereignty movements—demonstrate that technological alternatives serving human flourishing rather than corporate extraction are not only possible but already emerging through community organizing and democratic technology governance.

The author's analysis reveals the systematic nature of algorithmic oppression, but it also documents the systematic nature of community resistance. The question is not whether change is possible, but whether those of us with privilege and resources will use them in service of the transformation that the author's research shows is both necessary and achievable.

The evidence is overwhelming. The moral imperative is clear. The analytical frameworks exist. The time for action is now.

The author's research reveals that the future of technology is not predetermined by corporate interests or technological determinism. It will be shaped by the political choices we make and the power we build through collective action. Communities worldwide are demonstrating that another world is possible—one where technology serves human flourishing and environmental sustainability rather than profit maximization and social control.

The evidence the author has assembled points in only one direction: toward justice. The question is whether we will follow that evidence toward the transformation it demands, or continue accepting technological oppression as inevitable.

I know which side the evidence supports. The question is: which side are you on?

BIBLIOGRAPHY

African Union. (2022). *Continental data policy framework.* African Union Commission.

Albury, K. (2024). Algorithmic agency and "fighting back" against discriminatory Instagram content moderation: #IWantToSeeNyome. *Frontiers in Communication, 9,* Article 1385869. https://doi.org/10.3389/fcomm.2024.1385869

Alexander, M. (2010). The new Jim Crow: Mass incarceration in the age of colorblindness. The New Press.

American Civil Liberties Union of Maryland. (2021). The surveillance state of Baltimore: How the city's police surveillance network undermines community safety. ACLU of Maryland.

American Civil Liberties Union. (2023, August 6). *After third wrongful arrest, ACLU slams Detroit Police Department for continuing to use faulty facial recognition technology* [Press release]. https://www.aclu.org/press-releases/after-third-wrongful-arrest-aclu-slams-detroit-police-department-for-continuing-to-use-faulty-facial-recognition-technology

American Civil Liberties Union. (2024, June 28). *Civil rights advocates achieve the nation's strongest police department policy on facial recognition technology* [Press release]. https://www.aclu.org/press-releases/civil-rights-advocates-achieve-the-nations-strongest-police-department-policy-on-facial-recognition-technology

Angwin, J., Larson, J., Mattu, S., & Kirchner, L. (2016, May 23). Machine bias: There's software used across the country to predict future criminals. And it's biased against blacks. *ProPublica*. https://www.propublica.org/article/machine-bias-risk-assessments-in-criminal-sentencing

Barber, B. R. (1984). *Strong democracy: Participatory politics for a new age*. University of California Press.

Barlow, J. P., & Thorsen, E. (2021). Participatory technology assessment in comparative perspective: The cases of Denmark and the United States. *Technology in Society*, *64*, 101472. https://doi.org/10.1016/j.techsoc.2020.101472

Barocas, S., Hardt, M., & Narayanan, A. (2019). *Fairness and machine learning: Limitations and opportunities*. MIT Press.

Bartlett, R., Morse, A., Stanton, R., & Wallace, N. (2019). Consumer-lending discrimination in the FinTech era. *Journal of Financial Economics*, *143*(1), 30-56.

Baugh, J. (2018). *Linguistic profiling and discrimination*. Cambridge University Press.

Bauwens, M., Kostakis, V., & Pazaitis, A. (2019). *Peer to peer: The commons manifesto*. University of Westminster Press.

Benkler, Y. (2006). The wealth of networks: How social production transforms markets and freedom. Yale University Press.

Bianchi, A., & Flores, S. (2019). Community networks and local development: The case of Catalonia's Guifi.net. *Telecommunications Policy*, *43*(7), 101816. https://doi.org/10.1016/j.telpol.2019.04.001

Birrer, A., & Just, N. (2024). What we know and don't know about deepfakes: An investigation into the state of the research and regulatory landscape. *New Media & Society*, advance online publication. https://doi.org/10.1177/14614448241253138

Bolukbasi, T., Chang, K. W., Zou, J. Y., Saligrama, V., & Kalai, A. T. (2016). Man is to computer programmer as woman is to home-maker? Debiasing word embeddings. *Advances in Neural Information Processing Systems*, *29*, 4349-4357.

Bonilla, Y., & Rosa, J. (2015). #Ferguson: Digital protest, hashtag ethnography, and the racial politics of social media in the United States. *American Ethnologist*, *42*(1), 4-17. https://doi.org/10.1111/amet.12112

Booker, M. D. (2022). Empowering, engaging, and equipping technology through acceptance: A quantitative study utilizing the modified technology acceptance model (mTAM) to explore facial-recognition software acceptance (FRSA) in the smart policing initiative (SPI) [Doctoral dissertation, University of the Cumberlands]. ProQuest Dissertations & Theses Global.

Booker, M. D. (2025, February 18). Trump Administration's views on AI regulation is a step backward, EU AI act is a step forward. [Policy analysis].

Booker, M. D. (2025, February 19). Unraveling the DOGE data dilemma: Exploring the transformation to AI alchemy. [Unpublished manuscript].

Booker, M. D. (2025, June 20). The last text: How AI-generated images are driving teens to suicide. *LinkedIn*. [Article published by author]

Booker, M. D. (2025, March 17). Ph.D.: "Pretty Huge Discrepancies"--Examining African American Representation in Doctoral Programs in the United States of America. [Unpublished manuscript].

Brabenec, R. (2024, July 7). A billionaire, an AI supercomputer, toxic emissions and a Memphis community that did nothing wrong. *Tennessee Lookout.* https://tennesseelookout.com/2025/07/07/a-billionaire-an-ai-supercomputer-toxic-emissions-and-a-memphis-community-that-did-nothing-wrong/

Brantingham, P. J., Valasik, M., & Mohler, G. O. (2018). Does predictive policing lead to biased arrests? Results from a randomized controlled trial. *Statistics and Public Policy, 5*(1), 1-6.

Brookings Institution. (2023, June 27). Algorithmic bias detection and mitigation: Best practices and policies to reduce consumer harms. https://www.brookings.edu/articles/algorithmic-bias-detection-and-mitigation-best-practices-and-policies-to-reduce-consumer-harms/

Brown v. Board of Education, 347 U.S. 483 (1954).

Browne, S. (2015). Dark matters: On the surveillance of blackness. Duke University Press.

Bullard, R. D. (1990). Dumping in Dixie: Race, class, and environmental quality. Westview Press.

Buolamwini, J., & Gebru, T. (2018). Gender shades: Intersectional accuracy disparities in commercial gender classification. *Proceedings of Machine Learning Research, 81,* 77-91.

Butollo, F., & Krzywdzinski, M. (2018). Digitalization and the geographies of production: Towards reshoring or global fragmentation? *Competition & Change, 22*(4), 355-378. https://doi.org/10.1177/1024529418778320

Byrum, G. (2020). Municipal broadband: A solution to America's digital divide? *Government Finance Review, 36*(2), 18-23.

Cahn, A. F. (2021). The right to be seen: Algorithmic bias and civil rights. *Georgetown Law Technology Review, 5*(2), 412-456.

California Air Resources Board. (2023). *Bayview Hunters Point/Southeast San Francisco Community Emissions Reduction Plan.* https://ww2.arb.ca.gov/capp/com/cip/bayview-hunters-point-southeast-san-francisco

Coin ATM Radar. (2022). *Bitcoin ATM industry statistics and locations.* https://coinatmradar.com/charts/

Coleman, T. (2023, May 16). How algorithm discrimination affects social media. *The Week*. https://theweek.com/briefing/1023338/algorithm-ai-discrimination

Costanza-Chock, S. (2020). Design justice: Community-led practices to build the worlds we need. MIT Press.

Crawford, K. (2021). Atlas of AI: Power, politics, and the planetary costs of artificial intelligence. Yale University Press.

Crenshaw, K. (1989). Demarginalizing the intersection of race and sex: A black feminist critique of antidiscrimination doctrine, feminist theory and antiracist politics. *University of Chicago Legal Forum*, *1989*(1), 139-167.

Dastin, J. (2018, October 10). Amazon scraps secret AI recruiting tool that showed bias against women. *Reuters*. https://www.reuters.com/article/us-amazon-com-jobs-automation-insight/amazon-scraps-secret-ai-recruiting-tool-that-showed-bias-against-women-idUSKCN1MK08G

Dauvergne, P. (2022). Is artificial intelligence greening global supply chains? Exposing AI's real environmental impact. Polity Press.

Dencik, L., Hintz, A., Redden, J., & Treré, E. (2019). Exploring data justice: Conceptions, applications and directions. *Information, Communication & Society*, *22*(7), 873-881. https://doi.org/10.1080/1369118X.2019.1606268

Diakopoulos, N. (2016). Accountability in algorithmic decision making. *Communications of the ACM*, *59*(2), 56-62.

Dunbar-Hester, C. (2014). Low power to the people: Pirates, protest, and politics in FM radio activism. MIT Press.

Elliot, N., Klobucar, A., Breland, H. M., & Godley, A. J. (2012). Concerning validity: A response to the special issue on automated essay scoring. *Assessing Writing*, 17(4), 201-207. https://doi.org/10.1016/j.asw.2012.06.003

Environmental and Energy Study Institute. (n.d.). *Data centers and water consumption.* https://www.eesi.org/articles/view/data-centers-and-water-consumption

Estevez, E., Lopes, N., & Janowski, T. (2016). *Smart sustainable cities: Reconnaissance study.* United Nations University Operating Unit on Policy-Driven Electronic Governance.

Eubanks, V. (2018). Automating inequality: How high-tech tools profile, police, and punish the poor. St. Martin's Press.

European Commission. (2024). The Digital Services Act: Ensuring a safe and accountable online environment. Publications Office of the European Union.

European Commission. (n.d.). *The impact of the Digital Services Act on digital platforms.* https://digital-strategy.ec.europa.eu/en/policies/dsa-impact-platforms

European Parliament and Council. (2016). Regulation (EU) 2016/679 of the European Parliament and of the Council of 27 April 2016 on the protection of natural persons with regard to the processing of personal data and on the free movement of such data. *Official Journal of the European Union, L* 119, 1-88.

European Parliament and Council. (2024). Regulation (EU) 2024/1689 of the European Parliament and of the Council laying down harmonised rules on artificial intelligence. *Official Journal of the European Union, L* 1689, 1-144.

Florini, A. (2019). The coming democracy: New rules for running a new world. Island Press.

Flynn, A., Henry, N., Powell, A., & Scott, A. J. (2024). Non-consensual synthetic intimate imagery: Prevalence, attitudes, and knowledge in 10 countries. *Proceedings of the CHI Conference on Human Factors in Computing Systems.* https://doi.org/10.1145/3613904.3642382

Folley, A. (2020, September 22). Facebook suspends environmental groups, enraging activists. *The Hill.* https://thehill.com/

changing-america/sustainability/environment/517541-face-book-suspends-environmental-groups-enraging/

Freedom House. (2024). *The EU Digital Services Act: A win for transparency.* https://freedomhouse.org/article/eu-digital-services-act-win-transparency

Fung, A., & Warren, M. E. (2011). The participatory system: Democratic governance in the twenty-first century. In J. Fishkin & P. Laslett (Eds.), *Debating deliberative democracy* (pp. 17-33). Blackwell.

Gabrys, J. (2011). *Digital rubbish: A natural history of electronics.* University of Michigan Press.

Gangadharan, S. P., & Niklas, J. (2019). Decentering technology in discourse on discrimination. *Information, Communication & Society*, *22*(7), 882-899. https://doi.org/10.1080/1369118X.2019.1593484

Garcia, A., & Anderson, P. (2024). Elon Musk's xAI facility is using gas turbines in South Memphis, we're taking action. *Southern Environmental Law Center.* https://www.selc.org/news/resistance-against-elon-musks-xai-facility-in-south-memphis-gets-stronger/

Garcia, A., & Anderson, P. (2024, November 15). Elon Musk's xAI facility is polluting South Memphis. *Southern Environmental Law Center.* https://www.selc.org/news/elon-musks-xai-facility-is-polluting-south-memphis/

Garcia, E., & Weiss, E. (2020). COVID-19 and student performance, equity, and U.S. education policy. Economic Policy Institute.

Garvie, C., Bedoya, A., & Frankle, J. (2016). *The perpetual lineup: Unregulated police face recognition in America.* Georgetown Law Center on Privacy & Technology.

Gasman, M., Nguyen, T. H., Conrad, C. F., Lundberg, T., & Commodore, F. (2019). Black male doctoral students: Explaining influences and experiences that lead to degree completion. *Journal of Diversity in Higher Education, 12*(4), 300-311. https://doi.org/10.1037/dhe0000085

Ge, Y., Knittel, C. R., MacKenzie, D., & Zoepf, S. (2016). *Racial and gender discrimination in transportation network companies* (Working Paper No. 22776). National Bureau of Economic Research.

Gebru, T., Morgenstern, J., Vecchione, B., Vaughan, J. W., Wallach, H., Daumé III, H., & Crawford, K. (2021). Datasheets for datasets. *Communications of the ACM, 64*(12), 86-92. https://doi.org/10.1145/3458723

Gibson-Graham, J. K. (2006). *A postcapitalist politics.* University of Minnesota Press.

Gillborn, D., Warmington, P., & Demack, S. (2018). QuantCrit: Education, policy, 'Big Data' and principles for a critical race theory of statistics. *Race Ethnicity and Education, 21*(2), 158-179.

Global Witness. (2024, July 16). Facebook is the most toxic social media platform for climate activists – new survey suggests. https://globalwitness.org/en/press-releases/facebook-most-toxic-social-media-platform-climate-activists-new-survey-suggests/

Goodkind, N. (2021, August 24). Pregnant woman wrongfully arrested after facial recognition error. *Reuters.* https://www.reuters.com/world/us/pregnant-woman-wrongfully-arrested-after-facial-recognition-error-2021-08-24/

Government Accountability Office. (2023). Federal data centers: Agencies need to complete optimization efforts and improve reporting (GAO-23-105240).

Gradient Corporation. (2024, September 16). Environmental and community impacts of large data centers. https://gradient-corp.com/trend_articles/impacts-of-large-data-centers/

Grant, R. (2019). The extractive industries and society in the Global South. *The Extractive Industries and Society, 6*(2), 209-212. https://doi.org/10.1016/j.exis.2019.02.006

Green, B. (2019). The smart enough city: Putting technology in its place to reclaim our urban future. MIT Press.

Greenaction for Health and Environmental Justice. (2019). *Pollution, health, environmental racism and injustice: A toxic inventory of Bayview Hunters Point, San Francisco.* http://greenaction.org/wp-content/uploads/2019/06/thestate-oftheenvironment090204final.pdf

Grother, P., Ngan, M., & Hanaoka, K. (2019). *Face recognition vendor test (FRVT) part 3: Demographic effects* (NIST IR 8280). National Institute of Standards and Technology.

Gurstein, M. (2014). Smart cities vs. smart communities: Empowering citizens not market economics. *Journal of Community Informatics, 10*(3), 1-12.

Gurumurthy, A., & Chami, N. (2020). Digital decolonisation and the feminisation of development cooperation. *Development, 63*(2-4), 140-145. https://doi.org/10.1057/s41301-020-00267-z

Guskin, E. (2019, August 21). Flawed algorithms are grading millions of students' essays. *Vice.* https://www.vice.com/en/article/flawed-algorithms-are-grading-millions-of-students-essays/

Habermas, J. (1981). *The theory of communicative action* (T. McCarthy, Trans.). Beacon Press.

Hankivsky, O. (2014). Intersectionality-based policy analysis: An intersectional approach to health inequities. UBC Press.

Hanson, A., Hawkins, R. G., Sanchez, A., & Yinger, J. (2021, August 25). The secret bias hidden in mortgage-approval algorithms. *The Markup.* https://themarkup.org/denied/2021/08/25/the-secret-bias-hidden-in-mortgage-approval-algorithms

Hao, K. (2019, October 17). AI is sending people to jail—and getting it wrong. *MIT Technology Review.*

Haraway, D. (1988). Situated knowledges: The science question in feminism and the privilege of partial perspective. *Feminist Studies, 14*(3), 575-599. https://doi.org/10.2307/3178066

Harding, S. (1991). Whose science? Whose knowledge? Thinking from women's lives. Cornell University Press.

Harper, S. R. (2020). COVID-19 and the racial equity implications of reopening college and university campuses. *American Journal of Education, 127*(1), 153-162. https://doi.org/10.1086/711095

Hawkins, W., Russell, C., & Mittelstadt, B. (2025). Deepfakes on demand: The rise of accessible non-consensual deepfake image generators. *Proceedings of the Conference on Fairness, Accountability, and Transparency (FAccT '25)*. [Preprint]

Heilman, M., Cahill, A., Madnani, N., Lopez, M., Mulholland, M., & Tetreault, J. (2014). Predicting grammaticality on an ordinal scale. In *Proceedings of the 52nd Annual Meeting of the Association for Computational Linguistics* (pp. 174-180). Association for Computational Linguistics.

Henry, N., McGlynn, C., Flynn, A., Johnson, K., Powell, A., & Scott, A. J. (2020). Image-based sexual abuse: A study on the causes and consequences of non-consensual nude or sexual imagery. Routledge.

Hickel, J., Sullivan, D., & Zangwill, H. (2021). Rich countries drained $152tn from the global South since 1960. *Al Jazeera.* https://www.aljazeera.com/opinions/2021/5/6/rich-countries-drained-152tn-from-the-global-south-since-1960

Hill, K. (2020, January 18). Wrongfully accused by an algorithm. *The New York Times.* https://www.nytimes.com/2020/01/18/technology/clearview-privacy-facial-recognition.html

Hogan, M. (2015). Data flows and water woes: The Utah Data Center. *Big Data & Society, 2*(2), 1-12. https://doi.org/10.1177/2053951715592429

Hunt, P., Saunders, J., & Hollywood, J. S. (2016). Predictions put into practice: A quasi-experimental evaluation of Chicago's predictive policing pilot. RAND Corporation.

Hunter, R., Feltner, D., & Bourke, T. (2017). *Car insurance discrimination against low-income communities continues.* Consumer Federation of America.

Introna, L., & Wood, D. (2004). Picturing algorithmic surveillance: The politics of facial recognition systems. *Surveillance & Society*, *2*(2/3), 177-198.

Jahangir, R. (2025, April 5). Understanding the EU's Digital Services Act enforcement against X. *TechPolicy.Press.* https://www.techpolicy.press/understanding-the-eus-digital-services-act-enforcement-against-x/

Jasanoff, S. (2003). Technologies of humility: Citizen participation in governing science. *Minerva*, *41*(3), 223-244. https://doi.org/10.1023/A:1025557512320

Jefferson, B. J. (2020). Digitize and punish: Racial criminalization in the digital age. University of Minnesota Press.

Johnson, D. (2025, June 17). Elon Musk's xAI threatened with lawsuit over air pollution from Memphis data center, filed on behalf of NAACP. *NAACP.* https://naacp.org/articles/elon-musks-xai-threatened-lawsuit-over-air-pollution-memphis-data-center-filed-behalf

Jorgenson, A. K., & Clark, B. (2012). Are the economy and the environment decoupling? A comparative international study, 1960-2005. *American Journal of Sociology*, *118*(1), 1-44. https://doi.org/10.1086/665990

Kaba, M. (2021). We do this 'til we free us: Abolitionist organizing and transforming justice. Haymarket Books.

Katell, M., Young, M., Dailey, D., Herman, B., Guetler, V., Tam, A., Binz, C., Raz, D., & Krafft, P. M. (2020). Toward situated interventions for algorithmic equity: Lessons from the field. *Proceedings of the 2020 Conference on Fairness, Accountability, and Transparency*, 45-55. https://doi.org/10.1145/3351095.3372874

Kaye, D. (2019). Speech police: The global struggle to govern the internet. Columbia Global Reports.

Kee, K. F. (2017). Adoption and diffusion of broadband: A systematic review of the literature 2003-2015. *Telematics and Informatics*, *34*(8), 1456-1478. https://doi.org/10.1016/j.tele.2017.06.007

Kessler, S., Bach-Lombardo, N., & Haimson, J. (2022). *Hidden workers: Untapped talent*. Harvard Business School and Accenture.

Kiani, S. (2025, June 18). A historic Black community takes on the world's richest man over environmental racism. *Capital B News*. https://capitalbnews.org/musk-xai-memphis-black-neighborhood-pollution/

Knuth, S. (2022). Green colonialism in renewable energy. *Environment and Planning E: Nature and Space*, *5*(1), 47-71.

Larson, J., Mattu, S., Kirchner, L., & Angwin, J. (2016, May 23). How we analyzed the COMPAS recidivism algorithm. *ProPublica*. https://www.propublica.org/article/how-we-analyzed-the-compas-recidivism-algorithm

Love, B. L. (2019). We want to do more than survive: Abolitionist teaching and the pursuit of educational freedom. Beacon Press.

Luciano, K. (2021). Mining digital colonialism: How extractive industries affect local communities and ecosystems. *Third World Quarterly*, *42*(5), 1002-1020. https://doi.org/10.1080/01436597.2020.1870090

Lum, K., & Isaac, W. (2016). To predict and serve? *Significance*, *13*(5), 14-19.

Makarov, I., & Schoar, A. (2021). *Blockchain analysis of the Bitcoin market* (Working Paper No. 29396). National Bureau of Economic Research.

Marrinan, C. (2024, June 12). Data center boom risks health of already vulnerable communities. *TechPolicy.Press*. https://www.techpolicy.press/data-center-boom-risks-health-of-already-vulnerable-communities/

Masanet, E., Shehabi, A., Lei, N., Smith, S., & Koomey, J. (2020). Recalibrating global data center energy-use estimates. *Science*, *367*(6481), 984-986.

McClanahan, B., & South, N. (2021). The violence of security: State and corporate security and the commodification of every-day harm. *Theoretical Criminology*, *25*(2), 184-201.

Meijer, A., & Wessels, M. (2019). Predictive policing: Review of benefits and drawbacks. *International Journal of Public Adminis-tration*, *42*(12), 1031-1039.

Milan, S. (2013). Social movements and their technologies: Wiring social change. Palgrave Macmillan.

Milan, S., & Treré, E. (2019). Big data from the South(s): Beyond data universalism. *Television & New Media*, *20*(4), 319-335. https://doi.org/10.1177/1527476419837739

Mohamed, S., Png, M. T., & Isaac, W. (2020). Decolonial AI: Decolonial theory as sociotechnical foresight in artificial intelli-gence. *Philosophy & Technology*, *33*(4), 659-684.

Monserrate, S. G. (2022, February 22). The staggering ecological impacts of computation and the cloud. *The MIT Press Reader*. https://thereader.mitpress.mit.edu/the-staggering-ecological-im-pacts-of-computation-and-the-cloud/

Mytton, D. (2021). Data centre water consumption. *npj Clean Water*, *4*(1), 1-6.

Nanni, M. (2022). The carbon footprint of artificial intelligence. *Nature Machine Intelligence*, *4*(11), 968-970. https://doi.org/10.1038/s42256-022-00568-6

Narayan, A. (2023). The algorithmic auditing divide: How tech-nical and legal approaches to algorithmic accountability diverge. *Harvard Journal of Law & Technology*, *36*(2), 505-578.

National Center for Science and Engineering Statistics. (2024). *Doctorate recipients from U.S. universities: 2023* (NSF 25-300). U.S. National Science Foundation. https://ncses.nsf.gov/pubs/nsf25300

Ng, L. H. X., Cruickshank, I. J., & Carley, K. M. (2024). Does algorithmic content moderation promote democratic discourse? Radical democratic critique of toxic language AI. *Information, Communication & Society*, *27*(10), 1157-1176. https://doi.org/10.1080/1369118X.2024.2346531

Noble, S. U. (2018). Algorithms of oppression: How search engines reinforce racism. NYU Press.

O'Neil, C. (2016). Weapons of math destruction: How big data increases inequality and threatens democracy. Crown Publishers.

Oakland Privacy Advisory Commission. (2022). *Annual surveillance report*. City of Oakland. https://www.oaklandca.gov/topics/privacy-advisory-commission

Obermeyer, Z., Powers, B., Vogeli, C., & Mullainathan, S. (2019). Dissecting racial bias in an algorithm used to manage the health of populations. *Science*, *366*(6464), 447-453.

O'Neil, C. (2016). Weapons of math destruction: How big data increases inequality and threatens democracy. Crown Publishers.

Palfrey, J., & Gasser, U. (2016). Born digital: How children grow up in a digital age. Basic Books.

Parks, B. C., & Roberts, J. T. (2006). Globalization and vulnerability: The global political economy of food, energy and environment. Guilford Press.

Parks, L., & Starosielski, N. (Eds.). (2015). *Signal traffic: Critical studies of media infrastructures*. University of Illinois Press.

Pasek, A. (2020). Carbon futures: A infrastructure for climate uncertainty. *Cultural Studies*, *34*(6), 909-928. https://doi.org/10.1080/09502386.2019.1697127

Pasquale, F. (2015). The black box society: The secret algorithms that control money and information. Harvard University Press.

Patel, R., & Moore, J. W. (2017). A history of the world in seven cheap things: A guide to capitalism, nature, and the future of the planet. University of California Press.

Pater, J. A., Drouin, M., O'Connor, K., & Zytko, D. (2025). A commentary on sexting, sextortion, and generative AI: Risks, deception, and digital vulnerability. *Family Relations*, advance online publication. https://doi.org/10.1111/fare.13152

Pearson, K. (2025, June 26). Elon Musk's xAI project subjects Black communities in Memphis to more pollution. *MSNBC*. https://www.msnbc.com/opinion/msnbc-opinion/elon-musk-xai-naacp-memphis-rcna213745

Pellow, D. N. (2017). Total liberation: The power and promise of animal rights and the radical earth movement. University of Minnesota Press.

Pew Research Center. (2023). *The state of cryptocurrency adoption in America.* https://www.pewresearch.org/internet/2023/03/15/cryptocurrency-adoption/

Pohle, J., & Thiel, T. (2020). Digital sovereignty. *Internet Policy Review*, 9(4), 1-19. https://doi.org/10.14763/2020.4.1532

Polli, F. (2021, June 15). The future of fair hiring: Moving beyond bias in algorithms. *Harvard Business Review.*

Posselt, J. R., & Grodsky, E. (2017). Graduate education and social stratification. *Annual Review of Sociology*, *43*, 353-378. https://doi.org/10.1146/annurev-soc-081715-074324

Powell, A. (2021). Undoing optimization: Civic action in smart cities. Yale University Press.

Privacy International. (2017). *The global surveillance industry.* https://privacyinternational.org/sites/default/files/2017-12/global_surveillance_0.pdf

Raji, I. D., Smart, A., White, R. N., Mitchell, M., Gebru, T., Hutchinson, B., Smith-Loud, J., Theron, D., & Barnes, P. (2020). Closing the AI accountability gap: Defining an end-to-end framework for internal algorithmic auditing. *Proceedings of the 2020 Conference on Fairness, Accountability, and Transparency*, 33-44. https://doi.org/10.1145/3351095.3372873

Ramineni, C., & Williamson, D. M. (2018). Understanding mean score differences between the e-rater® automated scoring engine and humans for demographically based groups in the GRE® general test. *ETS Research Report Series*, 2018(1), 1-31. https://doi.org/10.1002/ets2.12192

Reardon, S. (2023). The CHIPS Act and American technological sovereignty: Progress and limitations. Brookings Institution.

Reich, J., & Mehta, J. (2020). Failure to disrupt: Why technology alone can't transform education. Harvard University Press.

Reigeluth, T. (2021). Why data is not enough: Digital maps and the production of ignorance. *Surveillance & Society*, *19*(4), 485-496. https://doi.org/10.24908/ss.v19i4.14170

Rey-Moreno, C., Roro, Z., Tucker, W. D., Siya, M. J., Bidwell, N. J., & Simo-Reigadas, J. (2013). Experiences, challenges and lessons from rolling out a rural WiFi-mesh network. *Proceedings of the 3rd ACM Symposium on Computing for Development*, 1-10. https://doi.org/10.1145/2442882.2442897

Ricaurte, P. (2019). Data epistemologies, the coloniality of knowledge, and resistance. *Television & New Media*, *20*(4), 350-365. https://doi.org/10.1177/1527476419831640

Richardson, R., Schultz, J., & Crawford, K. (2019). Dirty data, bad predictions: How civil rights violations impact police data, predictive policing systems, and justice. *New York University Law Review Online*, *94*, 15-55.

Roberts, D. (2011). Fatal invention: How science, politics, and big business re-create race in the twenty-first century. The New Press.

Roberts, J. T., & Parks, B. C. (2007). A climate of injustice: Global inequality, North-South politics, and climate policy. MIT Press.

Roberts, S. (2019). Behind the screen: Content moderation in the shadows of social media. Yale University Press.

Rodriguez, C. (2011). Citizens' media against armed conflict: Disrupting violence in Colombia. University of Minnesota Press.

Rosenblat, A. (2018). Uberland: How algorithms are rewriting the rules of work. University of California Press.

Rubel, A., & Jones, K. M. (2016). Student privacy in learning analytics: An information ethics perspective. *The Information Society*, *32*(2), 143-159.

Saltman, K. J. (2018). *The swindle of innovative educational finance*. University of Minnesota Press.

Sandvig, C. (2013). The internet as infrastructure and idea. In *Society and the internet: How networks of information and communication are changing our lives* (pp. 86-106). Oxford University Press.

Sandvig, C., Hamilton, K., Karahalios, K., & Langbort, C. (2014). Auditing algorithms: Research methods for detecting discrimination on internet platforms. *Data and Discrimination: Converting Critical Concerns into Productive Inquiry*, *22*, 4349-4357.

Scholz, T. (2016). Platform cooperativism: Challenging the corporate sharing economy. Rosa Luxemburg Foundation.

Selbst, A. D. (2021). An institutional view of algorithmic impact assessments. *Harvard Journal of Law & Technology*, *35*(1), 117-186.

Selbst, A. D., Boyd, D., Friedler, S. A., Venkatasubramanian, S., & Vertesi, J. (2019). Fairness and abstraction in sociotechnical systems. *Proceedings of the Conference on Fairness, Accountability, and Transparency*, 59-68. https://doi.org/10.1145/3287560.3287598

Settles, I. H., Jones, M. K., Buchanan, N. T., & Dotson, K. (2021). Epistemic exclusion: Scholar(ly) devaluation that marginalizes faculty of color. *Journal of Diversity in Higher Education*, *14*(4), 493-507. https://doi.org/10.1037/dhe0000174

Sklansky, D. A. (2018). The promise and perils of police body cameras. *University of Chicago Law Review*, *85*(5), 1125-1178.

Society for Human Resource Management. (2019). *The use of artificial intelligence in hiring*. https://www.shrm.org/hr-today/

trends-and-forecasting/research-and-surveys/pages/artificial-intelligence-hiring.aspx

Solnit, R. (2016). Hope in the dark: Untold histories, wild possibilities. Haymarket Books.

Southerland, V. (2022). The master's tools and a mission: Using community control and oversight laws to resist and abolish police surveillance technologies. *UCLA Law Review, 69*(4), 1018-1089.

Sovacool, B. K., & Hess, D. J. (2017). Ordering theories: Typologies and conceptual frameworks for sociotechnical change. *Social Studies of Science, 47*(5), 703-750.

Starosielski, N. (2015). *The undersea network.* Duke University Press.

Stroud, M. (2014, February 19). The minority report: Chicago's new police computer predicts crimes, but is it racist? *The Verge.* https://www.theverge.com/2014/2/19/5419854/the-minority-report-this-computer-predicts-crime-but-is-it-racist

Strubell, E., Ganesh, A., & McCallum, A. (2019). Energy and policy considerations for deep learning in NLP. *Proceedings of the 57th Annual Meeting of the Association for Computational Linguistics,* 3645-3650. https://doi.org/10.18653/v1/P19-1355

Sweeney, L. (2013). Discrimination in online ad delivery. *Communications of the ACM, 56*(5), 44-54.

Tanner, A. (2014). The learning machines: How big data has transformed education, work, and the search for human potential. Basic Books.

Te Mana Raraunga. (2018). *Māori data sovereignty principles.* Te Mana Raraunga - Māori Data Sovereignty Network.

Transnational Institute. (2021). *The energy transition in North Africa.* https://www.tni.org/en/article/the-energy-transition-in-north-africa

Tuck, E., & Yang, K. W. (2012). Decolonization is not a metaphor. *Decolonization: Indigeneity, Education & Society, 1*(1), 1-40.

Tufekci, Z. (2017). Twitter and tear gas: The power and fragility of networked protest. Yale University Press.

Tuttle, B. T., & Baxstrom, R. (2021). Community-based violence interruption: A public health approach to reducing gun violence. *Annual Review of Public Health*, *42*, 289-307.

U.S. Congressional Budget Office. (2023). The federal government's use of artificial intelligence: Benefits, challenges, and potential policy responses (CBO Publication 58946).

Umbach, R., Henry, N., Beard, G., & Berryessa, C. (2024). "Violation of my body:" Perceptions of AI-generated non-consensual (intimate) imagery. [Preprint] https://arxiv.org/html/2406.05520v1

UN Office of the High Commissioner for Human Rights. (2024, July). Racism and AI: "Bias from the past leads to bias in the future." https://www.ohchr.org/en/stories/2024/07/racism-and-ai-bias-past-leads-bias-future

Vallejos, R. (2024, June 5). Data centers bring environmental concerns, like excess water use, to Chile. *Rest of World*. https://restofworld.org/2024/data-centers-environmental-issues/

Van Brakel, R., & De Hert, P. (2011). Policing, surveillance and law in a pre-crime society: Understanding the consequences of technology based strategies. *Journal of Police Studies*, *20*(3), 163-192.

Vidal, O., Le Boulzec, H., & François, C. (2018). The impact of the end of Moore's law on the environmental sustainability of the ICT sector. *Sustainability*, *10*(12), 4678. https://doi.org/10.3390/su10124678

Wachter, S., Mittelstadt, B., & Floridi, L. (2017). Why a right to explanation of automated decision-making does not exist in the general data protection regulation. *International Data Privacy Law*, *7*(2), 76-99. https://doi.org/10.1093/idpl/ipx005

Wang, X., Wu, Y. C., Ji, X., & Fu, H. (2024). Algorithmic discrimination: Examining its types and regulatory measures with emphasis on US legal practices. *Frontiers in Artificial Intelligence, 7*, Article 1320277. https://doi.org/10.3389/frai.2024.1320277

Washington, A. L. (2018). How to argue with an algorithm: Lessons from the COMPAS-ProPublica debate. *Colorado Technology Law Journal, 17*(1), 131-160.

Watters, A. (2021). Teaching machines: The history of personalized learning. MIT Press.

Williamson, B. (2017). Big data in education: The digital future of learning, policy and practice. SAGE Publications.

Wilson, S., Hoffman, M., Morgenstern, J., & Koenecke, A. (2021). Predictive inequity in object detection. *Proceedings of the 2021 AAAI/ACM Conference on AI, Ethics, and Society*, 954-963.

Winner, L. (1980). Do artifacts have politics? *Daedalus, 109*(1), 121-136.

Winner, L. (1985). Do artifacts have politics? In D. MacKenzie & J. Wajcman (Eds.), *The social shaping of technology* (pp. 26-38). Open University Press.

Wittes, B., & Poplin, C. (2016). *Sextortion: Cybersecurity, teenagers, and remote sexual assault.* Brookings Institution Center for Technology Innovation. https://www.brookings.edu/blog/techtank/2016/05/11/sextortion-the-problem-and-solutions/

World Bank & International Telecommunication Union. (2024). *Green data centers: Towards a sustainable digital transformation.* https://www.worldbank.org/en/topic/digital/publication/green-data-centers-towards-a-sustainable-digital-transformation

Wu, C. C. (2019). Past imperfect: How credit scores and other analytics "bake in" and perpetuate past discrimination. National Consumer Law Center.

Young, W. (2019). The invisible tax on Black students. *The Hechinger Report.* https://hechingerreport.org/the-invisible-tax-on-black-students/

Zuboff, S. (2019). The age of surveillance capitalism: The fight for a human future at the new frontier of power. PublicAffairs.

INDEX

Sanchez, David, 107

Science, Technology, and Society (STS), 173

Shadow-banning, 131

Smart contracts, 98

Social media algorithms, 128, 130-131

Social Security Administration algorithms, 114-115

South Korea Digital New Deal, 205

South Memphis, 56, 58-60, 71, 218

Spatial displacement, 261-262

Stocksy United cooperative, 301, 316

Strategic Subject List, 20

Supply chain environmental racism, 262-263

Surveillance infrastructure, 41-55

Sustainable development, 264-266, 330-331

Sweden environmental assessment, 203, 255

Sweeney, Latanya, 14

T

Technology assessment, 175, 181

Technology cooperatives, 301-302

Technology education, 232-233

Technology governance, 172-194, 272-296

Technology transfer, 280, 323-324

Title VI Civil Rights Act, 177

Traditional ecological knowledge, 184, 258

Transparency mechanisms, 219-223, 283-284

Treaty of Waitangi, 212

U

Uber algorithms, 100-101

University of Texas algorithmic screening, 147

Urban environmental justice, 41-55

V

Vallejos, Rodrigo, 83

Video interview analysis, 36, 91

W

X

Z

ABOUT THE AUTHOR

———————————

D r. Mario DeSean Booker, Ph.D. bridges technical expertise with community advocacy, connecting rigorous analysis to social justice organizing. His work demonstrates how technological research can serve community empowerment while maintaining academic credibility.

Early Life and Education

Dr. Mario Desean Booker is a professor of Information Technology at Purdue University Global, where he teaches graduate-level courses in computer networks, advanced network management, information security, and cloud operations. He has also served as faculty and subject matter expert at multiple universities across the United States—including the University of Michigan–Flint, Trine University, Salem University, and Cleary University—designing and delivering curricula in cybersecurity, data science, cloud computing, and digital forensics. His teaching portfolio spans undergraduate to doctoral levels, with contributions in areas such as network security, penetration testing, virtualization, predictive analytics, and emerging cloud architectures.

Dr. Booker holds a Ph.D. in Information Technology from the University of the Cumberlands with a specialization in digital and network forensics, an M.S. in Technology and Innovation Management (specialization in data science), and dual B.A.S. degrees in Information Systems Engineering and Networking & Security. He has further distinguished himself with industry-recognized certifications, including CompTIA A+, AWS Cloud Practitioner, and ServiceNow IT Leadership Professional, as well as Quality Matters (QM) Higher Education Peer Reviewer credentials.

An active researcher and thought leader, Dr. Booker has published extensively on topics including blockchain technologies, big data analytics, ransomware, and the ethical implications of information systems. His scholarship is complemented by professional experience in systems management, network administration, and IT service delivery, notably with the State of Michigan Department of Health and Human Services and Fortune 500 corporations. He contributes to the academic community through peer reviewing, curriculum development, and active involvement in research dissemination via ResearchGate and various international journals.

Dr. Booker's research lies at the intersection of emerging technologies, cybersecurity, and technological governance, with a strong interdisciplinary foundation that draws from computer science, science and technology studies (STS), critical race theory, environmental justice, and public policy. His early work investigated biometric and IoT technologies in smart policing, identifying ethical and operational gaps in facial recognition, gunshot detection, and biometric tracking systems. Building on this foundation, his scholarship has evolved into a comprehensive examination of AI ethics, cybersecurity bias, environmental justice, platform governance, and financial technologies.

Beyond academia, Dr. Booker engages in public service and STEM outreach, having served as a science fair judge for the Regeneron International Science and Engineering Fair, the Flint Regional Science and Engineering Fair, and Broadcom MASTERS. He has also contributed to community advisory boards addressing technology, equity, and public safety.

Through his teaching, research, and professional service, Dr. Booker advances both the scholarly discourse and applied practice of information technology, with particular emphasis on the intersections of cybersecurity, data science, and cloud computing.

Academic Foundation and Technical Expertise

Dr. Booker earned his Ph.D. in Information Technology with specialization in Digital and Network Forensics from the University of the Cumberlands, graduating Magna Summa Cum Laude. His technical credentials include dual bachelor's degrees in Information Systems Engineering and Cybersecurity, plus a Master of Science in Technology and Innovation Management with Data Science specialization.

Currently Full-Time Professor of Graduate Information Technology at Purdue University Global, he has held academic appointments at the University of Michigan-Flint, Trine University, and Salem University. Dr. Booker teaches courses spanning network security, cloud computing, digital forensics, and data visualization, connecting technical concepts to real-world social impacts.

Research Innovation and Scholarly Impact

Dr. Booker's research evolved from foundational biometric systems analysis to pioneering work on digital colonialism and environmental racism in technology infrastructure. His doctoral dissertation utilized the Modified Technology Acceptance Model to examine facial recognition software acceptance in smart policing initiatives.

His current research introduces novel analytical frameworks quantifying bias-driven cybersecurity vulnerabilities through the Intersectional Disparity Index (IDI) and Cumulative Disadvantage Score (CDS). These metrics reveal how demographic performance disparities in AI-driven security systems create exploitable attack surfaces, bridging cybersecurity with social justice analysis.

Dr. Booker's 2025 publications span environmental justice ("Digital Redlining: AI Infrastructure and Environmental Racism"), regulatory analysis ("CTRL+ALT+DELETE: Congress Attempts to Reboot AI Regulation"), and financial technology ("Currency of Change: Encrypting the Future of Monetary Governance"), demonstrating intellectual range across technology policy domains.

Community Leadership and Practical Application

Dr. Booker served on the Flint Community Advisory Taskforce for Public Safety (2020-2022), translating technical knowledge into community empowerment strategies. This role involved addressing systemic racism in policing while developing "Truth, Racial Healing, and Transformation" training programs bridging academic analysis with grassroots organizing.

As co-minister of HOPE (Helping Other People Elevate) Outreach Ministries in Flint, he addresses urban poverty, violence, and environmental degradation through community problem-solving. His Doctorate of Divinity with New Testament Gospels specialization provides theological grounding for social justice work while informing understanding of technology's moral dimensions.

His 2020 Flint Community Schools Board campaign demonstrated commitment to educational equity and community self-determination.

Theoretical Innovation and Global Perspective

Dr. Booker develops theoretical frameworks connecting local community impacts to global technological exploitation patterns. His digital colonialism analysis reveals how AI infrastructure deployment functions as environmental warfare against marginalized communities, while environmental racism research documents systematic corporate withdrawal from white communities.

His Memphis case study of Elon Musk's xAI "Colossus" supercomputer facility exemplifies his analytical approach, connecting corporate behavior, environmental health impacts, and community resistance within broader technological governance patterns. This methodology demonstrates how individual incidents reveal systemic patterns while making abstract concepts accessible to organizers and policymakers.

His international perspective encompasses comparative policy analysis across continents, examining how nations handle technology infrastructure sovereignty and environmental justice. Work on "server farm colonialism" in Chile and "data colonialism" in Kenya reveals global dimensions of technological environmental racism while identifying successful resistance models.

Methodological Contributions

Dr. Booker's methodology integrates quantitative technical analysis with qualitative community-based research, creating interdisciplinary approaches serving both academic rigor and community empowerment. His community-based algorithmic auditing approaches provide tools for grassroots organizations challenging discriminatory systems.

Future research directions include comprehensive AI ethics framework development, climate-conscious technology governance, and international technology policy coordination, building on established expertise while addressing emerging AI governance and digital rights challenges.

Personal Background and Values

Born and raised in Flint, Michigan, Dr. Booker's experience with environmental racism through the water crisis informs his understanding of systematic institutional harm and community resistance strategies. Married to FaLessia for 19 years with two sons, Jaiden and Jace, his family commitments ground academic work in lived community experience.

Distinctive Contributions

Dr. Booker's work makes several unique contributions to technology governance:

Environmental Justice Integration: First comprehensive analysis connecting algorithmic discrimination to environmental racism, revealing how AI infrastructure deployment perpetuates historical inequities.

Digital Colonialism Framework: Novel theoretical approach understanding infrastructure imperialism in the AI era with concrete policy applications.

Community-Centered Methodology: Integration of organizing experience with academic research creating scholarship serving both intellectual advancement and grassroots empowerment.

Interdisciplinary Synthesis: Successful integration of computer science, environmental justice, critical race theory, and international relations creating new analytical possibilities.

Dr. Booker demonstrates that rigorous technical analysis and community advocacy reinforce each other, creating engaged scholarship serving both intellectual development and social transformation. His work provides essential contributions to understanding how emerging technologies can challenge existing inequalities.